Topics in Recreational Mathematics 1/2016

Editor-in-chief

Charles Ashbacher
5530 Kacena Ave
Marion, IA 52302 USA

cashbacher@yahoo.com

Artwork

Caytie Ribble

Problems

Lamarr Widmer

ISBN-13: 978-1530004225

CONTENTS

Note from the Editor

Welcome to the latest iteration of the **Topics in Recreational Mathematics** series. This is volume 6 and the first issue of 2016. Last year was an amazing one, when **Journal of Recreational Mathematics** ceased publication, the decision was made to keep recreational mathematics going with a series of books.

The initial thought was to publish all of the material that had been accepted for publication in **Journal of Recreational Mathematics** so that it would not be lost. That task has been accomplished and now new material is being published. I am grateful to the people that have and continue to contribute their work for publication.

Humans love to solve basic puzzles and many of them have a foundation in mathematics. Recreational mathematics has done a great deal to satisfy that innate human desire and it is hoped that this series plays a part in this continuing endeavor. I welcome submissions for future issues to the email address below.

As always, I welcome feedback and comments.

Charles Ashbacher

cashbacher@yahoo.com

5530 Kacena Ave

Marion, IA 52302

Mathematical Cartoons

Caytie Ribble

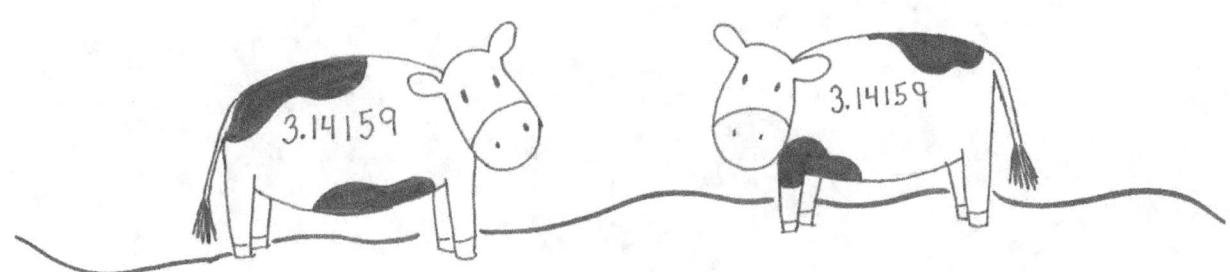

DIPOLE

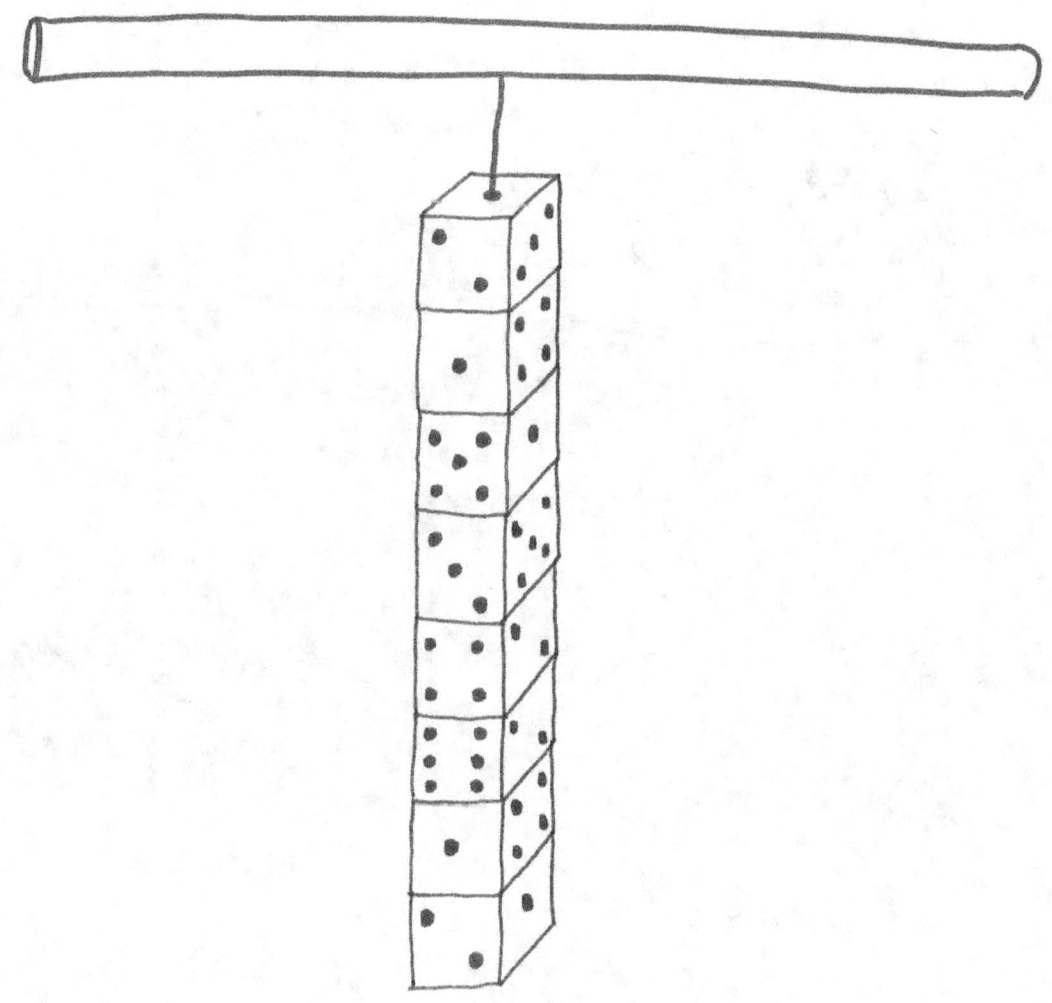

The Impact of Free Agency on NHL Player Performance

Emily A. Fluke

Paul M. Sommers

Department of Economics
Middlebury College
Middlebury, Vermont 05753
psommers@middlebury.edu

Abstract

The authors use "point shares" to assess the impact of free agency on National Hockey League player performance by dividing restricted and unrestricted free agents in 2013-14 into two groups, players who stayed with their same team and those who switched to another team. Restricted free agents (and to a lesser extent, unrestricted free agents) who play for a new team perform worse and those who play for their previous team perform marginally better. The study underscores the importance of (i) separating restricted from unrestricted free agents and (ii) noting that staying with the same team or playing for a new one differentially affects free agent performance.

The new collective bargaining agreement that emerged from the National Hockey League's 301-day lockout in 2004-2005 (and ratified again in 2012-13) described the myriad ways when and how a player can become a free agent. In general, an NHL player whose contract has expired can become an unrestricted free agent (hereafter UFA) once he has reached age 27 or has seven years of service.[1,2] A player under the age of 25 whose contract has expired and has three years of NHL experience can become a restricted free agent (hereafter RFA). UFAs are free to sign with any team that makes them an offer; RFAs are free to solicit offers from other teams, but their original team has the "right of first refusal," meaning it can retain the player by matching the other team's offer [1, p. 301]. In general, the category of free agency in the NHL depends on the player's age and level of experience.

A new metric called "point shares" (developed by Justin Kubatko) is an estimate of the number of points (the number of goals plus assists) contributed by an NHL player.

Negative point shares are possible, if the player's on-ice performance is so egregious that it takes away points that his teammates generated. There is a defensive component as well as an offensive component which are summed for each player and reported by www.hockey-reference.com among the NHL player's "miscellaneous" statistics (see www.hockey-reference.com/about/point_shares.html for a more detailed description of point shares).

In this brief note, we compare point shares of the NHL's 2013-14 free agents in the year they became free agents and the year after. For each group of free agents (UFAs and RFAs), we divide the sample into two additional groups: (i) players who signed with a new team and (ii) players who signed with their previous team. We examine defensemen and forwards separately, but not goaltenders. Moreover, free agents who went to arbitration are excluded from our sample as are players who did not play in the strike-shortened 2012-13 season before they became free agents in 2013-14. Our sample includes 68 RFAs (29 defensemen and 39 forwards) and 101 UFAs (31 defensemen and 70 forwards) who played all three consecutive seasons (2012-13, 2013-14, and 2014-15).

Table 1 shows the average point shares for each of the four groups of free agents, as well as for all RFAs and UFAs. At first glance, average point shares for each group do not change from 2013-14 to 2014-15. In each of the six paired t-tests, the p-value is not small enough to convincingly rule out chance. That is, for each comparison, there is no statistically significant difference between the two average point shares (using $\alpha = .05$).

Now let's make a distinction between free agents who sign with a new team and those who sign with their previous team.[3] If a free agent signs with a new team, his performance may suffer as he tries to deal with the stress of relocating and acclimating to new teammates. Forming forward lines and defensive pairs is a strategic process, and players who can spend more time learning how their teammates think and play should be able to perform better than players with brand new teammates. As for free agents who re-sign with their previous team, such an expression of approval and encouragement from management may inspire these grateful players to perform even better than they did the year they filed for free agency.

Table 2 shows the results of eight paired t-tests to gauge the performance of the same players in the year they became free agents and the year after. RFA defensemen who signed with a new team performed significantly worse ($p = .023$) and those who signed with their previous team performed marginally better ($p = .096$). RFA forwards who signed with their previous team also performed marginally better ($p = .076$). The average performance of the more experienced UFA defensemen did not change from one year to the next, no matter who they signed with. UFA forwards (like RFA defensemen) performed marginally worse when they signed with a new team ($p = .083$).

In Table 3, comparisons are made of the average incremental change in point shares for the four different groups of free agents when they (i) signed with a new team or (ii) signed with their previous team. The last section of this table underscores the earlier finding that less experienced RFAs (either defensemen or forwards) who signed with their previous team performed much better than their counterparts who signed with a new team.

Concluding Remarks

An analysis of the NHL's 2013-14 free agents reveals that restricted free agents (notably defensemen) who sign with a new team perform significantly worse in the year after free agency, while restricted free agents (both defensemen and forwards) who sign with their previous team perform marginally better in the year after free agency. Unrestricted free agents (notably forwards) who sign with a new team also perform marginally worse.

In general, the results show how performance in the year after free agency depends on whether the player remained with his team or switched to another team. And, in particular, teams stand to benefit if they can find a way to reach an agreement to a contract extension with their restricted free agents.

Table 1. Average Point Shares of 2013-14 Free Agents

in 2013-14 and 2014-15

	2013-14	2014-15	p-value on difference*
Defensemen			
RFAs (n = 29)	3.324	3.393	.863
UFAs (n = 31)	2.545	2.313	.537
Forwards			
RFAs (n = 39)	2.033	2.185	.465
UFAs (n = 70)	1.841	1.563	.189
Defensemen and Forwards			
RFAs (n = 68)	2.584	2.700	.571
UFAs (n = 101)	2.057	1.793	.153

*All p-values are reported for a two-tailed t-test.

Table 2. Results of Paired *t*-Tests

for 2013-14 Free Agents

	Signed with new team			Signed with previous team		
	Average		*p*-value	Average		*p*-value
	2013-14 point share	2014-15 point share	on difference	2013-14 point share	2014-15 point share	on difference*
Defensemen						
RFAs	3.329	1.414	**.023** (n = 7)	3.323	4.023	*.096* (n = 22)
UFAs	2.580	2.400	.690 (n = 25)	2.400	1.950	.453 (n = 6)
Forwards						
RFAs	0.762	0.300	.165 (n = 13)	2.669	3.127	*.076* (n = 26)
UFAs	2.053	1.635	*.083* (n = 57)	0.915	1.246	.448 (n = 13)

*All *p*-values are reported for a two-tailed *t*-test.

Table 3. Incremental Change in Point Shares
from 2013-14 to 2014-15,
between groups of free agents

	Average of Group 1	Group 2	Average of Group 2	*p*-value on difference*
Group 1				

Both groups signed with a new team in 2014-15

	Average of Group 1	Group 2	Average of Group 2	*p*-value on difference*
Defensemen, RFAs	-1.914	Defensemen, UFAs	-0.180	*.066*
Defensemen, RFAs	-1.914	Forwards, RFAs	-0.462	**.032**
Defensemen, RFAs	-1.914	Forwards, UFAs	-0.418	**.040**
Defensemen, UFAs	-0.180	Forwards, RFAs	-0.462	.672
Defensemen, UFAs	-0.180	Forwards, UFAs	-0.418	.609
Forwards, RFAs	-0.462	Forwards, UFAs	-0.418	.933

Table 3. Incremental Change in Point Shares

from 2013-14 to 2014-15,

between groups of free agents

(Continued)

Group 1	Average of Group 1	Group 2	Average of Group 2	p-value on difference*

Both groups signed with their previous team in 2014-15

Group 1	Average of Group 1	Group 2	Average of Group 2	p-value on difference*
Defensemen, RFAs	0.700	Defensemen, UFAs	-0.450	.176
Defensemen, RFAs	0.700	Forwards, RFAs	0.458	.598
Defensemen, RFAs	0.700	Forwards, UFAs	0.331	.553
Defensemen, UFAs	-0.450	Forwards, RFAs	0.458	.127
Defensemen, UFAs	-0.450	Forwards, UFAs	0.331	.298
Forwards, RFAs	0.458	Forwards, UFAs	0.331	.784

*All p-values are reported for a two-tailed t-test.

Table 3. Incremental Change in Point Shares
from 2013-14 to 2014-15,
between groups of free agents
(Continued)

| | Average of | | Average of | *p*-value on |
Group 1	Group 1	Group 2	Group 2	difference*

Group 1 signed with their previous team in 2014-15

Group 2 signed with a new team in 2014-15

Defensemen,		Defensemen,		
RFAs	0.700	RFAs	-1.914	**.003**
Forwards,		Forwards,		
RFAs	0.458	RFAs	-0.462	**.033**
Defensemen,		Defensemen,		
UFAs	-0.450	UFAs	-0.180	.780
Forwards,		Forwards,		
UFAs	0.331	UFAs	-0.418	.167

*All *p*-values are reported for a two-tailed *t*-test.

Reference

1 .   Leeds, Michael A. and Peter von Allmen. *The Economics of Sports*,

Pearson Education, Inc., 2014

Footnotes

1.   Players 25 years of age or older, who have completed three or more seasons
of professional hockey and have played fewer than 80 NHL (regular season
and playoff) games, may also qualify as unrestricted free agents.

2.   To put the 27 or 7 rule for unrestricted free agency into perspective,
according to www.quanthockey.com/Distributions/CareerLengthGP.php
average NHL career length is 5.65 seasons. The average NHL player
retires at age 28.

3 .   If a player was traded part way through the 2014-15 season, we considered
him on a new team after his free agency year and used only his point shares
with his new team.

Some Conjectures on the Carmichael Numbers

Marius Coman

Charles Ashbacher

Abstract

In his book "Two Hundred Conjectures and One Hundred and Fifty Open Problems On Fermat Pseudoprimes" Marius Coman states several conjectures about the Carmichael numbers. The purpose of this short paper is to state some of them that seem amenable to computer analysis.

The book **Some Conjectures on the Carmichael Numbers** [1] by Marius Coman contains many conjectures and unsolved problems. The purpose of this short note is to list some the conjectures regarding the Carmichael numbers that can be analyzed using basic computer techniques.

Definition: A Carmichael number is a composite number n that satisfies the congruence relation

$$b^{n-1} \equiv 1 \ (mod \ n)$$

for all b, such that $1 < b < n$.

Conjecture: The number $(30 * n + 7) * (60 * n + 13) * (150 * n + 31)$ is a Carmichael number if (but not only if) $30 * n + 7$, $60 * n + 13$ and $150 * n + 31$ are all prime numbers.

Conjecture: The number

$(30 * n - 29) * (60 * n - 59) * (90 * n - 89) * (180 * n - 179)$

is a Carmichael number if (but not only if) $30 * n - 29$, $60 * n - 59$, $90 * n - 89$ and $180 * n - 179$ are all prime numbers.

Conjecture: The number

$(330 * n + 7) * (660 * n + 13) * (990 * n + 19) * (1980 * n + 37)$

is a Carmichael number if $330 * n + 7$, $660 * n + 13$, $990 * n + 19$ and $1980 * n + 37$ are all prime numbers.

Conjecture: The number $(30 * n - 7) * (90 * n - 23) * (300 * n - 79)$ is a Carmichael number if (but not only if) $30 * n - 7$, $90 * n - 23$ and $300 * n - 79$ are all prime numbers.

Conjecture: The number $(30 * n + 13) * (90 * n + 37) * (150 * n + 61)$ is a Carmichael number if (but not only if) $30 * n + 13$, $90 * n + 37$ and $150 * n + 61$.

Reference

1. Coman, Marius. *Some Conjectures on the Carmichael Numbers*, Education Publishing, Columbus, Ohio, 2003. ISBN 978-1-59973-257-2.

Linear Algebra Properties of Magic Squares

Hossein Behforooz
Mathematics Department
Utica College
Utica, New York 13502
hbehforooz@utica.edu
www.utica.edu/hbehforooz

Abstract

Overall, every magic square is a very special square matrix and in this article we are going to show some interesting linear algebra properties of these magic square matrices. There are some published short articles on this subject but they are not complete papers with all properties in one article.

Introduction

In this paper, we will use the familiar matrix notations and matrix operations for our studies. For example, $A = [a_{i,j}]$ means a magic square of order n with entries $a_{i,j}$ when i, j=1, 2, ... n. As we know, in every magic square, the sum of all rows and columns and two diagonals are equal to one constant number S which is called the magic sum. If the entries of each diagonal add up to a number different than S then A is called a semi-magic square. Now, we will state a few theorems on the linear algebra properties of magic square matrices. Since we are in the area of recreational mathematics, we will demonstrate and present the proofs by simple examples and leave the actual mathematical proofs to the interested readers. I have tried and investigated with additional examples and these properties are true for other magic squares as well.

Theorem 1: If $A = [a_{i,j}]$ and $B = [b_{i,j}]$ are two n by n magic squares with magic sums S_1 and S_2 then (with matrix addition, subtraction and scalar multiplication operations) the linear combination $\alpha A + \beta B$ is a magic square with magic sum equal to $\alpha S_1 + \beta S_2$, where α and β are two arbitrary real numbers.

The proof is easy and obvious. This theorem shows that we can add or subtract two magic squares A and B and obtain other magic squares with magic sums equal to $S_1 + S_2$ or $S_1 - S_2$.

Theorem 2: If $A = [a_{i,j}]$ and $B = [b_{i,j}]$ are two n by n magic squares with magic sums S_1 and S_2 then (with matrix multiplication operation) the product AB is a magic square or semi-magic square with magic sum equal to $S_1 S_2$.

Demonstration: Consider the following two 3 by 3 magic squares A and B with magic sums $S_1 = 15$ and $S_2 = 42$. Then we see that AB is a semi-magic square with magic sum equal to $S = 630 = 15 \times 42$ (you try other examples!).

$$A = \begin{bmatrix} 4 & 9 & 2 \\ 3 & 5 & 7 \\ 8 & 1 & 6 \end{bmatrix} \qquad B = \begin{bmatrix} 13 & 8 & 21 \\ 22 & 14 & 6 \\ 7 & 20 & 15 \end{bmatrix} \qquad AB = \begin{bmatrix} 264 & 198 & 168 \\ 198 & 234 & 198 \\ 168 & 198 & 264 \end{bmatrix}$$

Theorem 3: If $A = [a_{i,j}]$ is a magic square with magic sum S then $A^2, A^3, A^4, \ldots$ are magic squares or semi-magic squares with magic sums equal to $S^2, S^3, S^4, \ldots$

Demonstration: Consider the previous magic square matrix A with magic sum $S = 15$. If we compute the matrices A^2, A^3, A^4, A^5, then we easily see that these are a semi-magic square, magic square, semi-magic square and magic square with magic sums equal to:

$$S^2 = 225 = 15^2, S^3 = 3375 = 15^3, S^4 = 50625 = 15^4, S^5 = 759375 = 15^5.$$

(You try other examples). We can prove that all of the even exponents are semi-magic squares and the odd exponents are magic squares, see the comments in [3, page 61].

$$A = \begin{bmatrix} 4 & 9 & 2 \\ 3 & 5 & 7 \\ 8 & 1 & 6 \end{bmatrix}, \quad A^2 = \begin{bmatrix} 59 & 83 & 83 \\ 83 & 59 & 83 \\ 83 & 83 & 59 \end{bmatrix}, \quad A^3 = \begin{bmatrix} 1149 & 1029 & 1197 \\ 1173 & 1125 & 1077 \\ 1053 & 1221 & 1101 \end{bmatrix},$$

$$A^4 = \begin{bmatrix} 17259 & 16683 & 16683 \\ 16683 & 17259 & 16683 \\ 16683 & 16683 & 17259 \end{bmatrix}, \quad A^5 = \begin{bmatrix} 252549 & 255429 & 251397 \\ 251973 & 253125 & 254277 \\ 254853 & 250821 & 253701 \end{bmatrix}.$$

Interestingly, if we look more closely, we notice that the first two digits of the entries in A^4 are the entries of A^2 and the other three digits in A^4 are (except the diagonal entries) twice of those two digits. Similar properties are between the entries of A^6 and A^2. Are these coincidences?

Now we will examine the inverses of magic square matrices. Some magic square matrices are non-singular, invertible and their inverses are also magic squares. But there are some singular magic squares without inverse matrices.

Theorem 4: The inverse matrix of a non-singular magic square matrix is a magic square with magic sum equal to the reciprocal of the magic sum of the original magic square.

Demonstration: Consider the magic square A with magic sum S = 15.

In this example, the det (A) =360=15×24. Since the determinant value is non-zero, its inverse matrix A^{-1} exists. In general, we have a relation between the inverse of A, det (A) and its adjoint matrix adj (A), which is

$$A^{-1} = \frac{adj(A)}{det(A)}.$$

If in our example, we find the adj (A) then the inverse of A will be:

$$A^{-1} = \frac{1}{360} \begin{bmatrix} 23 & -52 & 53 \\ 38 & 8 & -22 \\ -37 & 68 & -7 \end{bmatrix}.$$

As you see, the above adjoint matrix is a magic square with magic sum equal to 24 and also A^{-1} is a magic square with magic sum equal to 1/15, see also [2] and [4].

We have a very interesting property between the magic sums and the eigenvalues of the magic squares.

Theorem 5: The magic sum of every magic square matrix is the eigenvalue of the magic square matrix.

Demonstration: Consider the magic square A with magic sum $S = 15$.

The eigenvalue of this matrix is the solution λ of the following determinant equation or characteristic polynomial:

$$A = \begin{bmatrix} 4 & 9 & 2 \\ 3 & 5 & 7 \\ 8 & 1 & 6 \end{bmatrix}, \quad \det(A - \lambda I) = \det \begin{bmatrix} 4-\lambda & 9 & 2 \\ 3 & 5-\lambda & 7 \\ 8 & 1 & 6-\lambda \end{bmatrix} = 0.$$

If we expand this determinant and do a little bit algebra we will get:

$$\det(A - \lambda I) = -\lambda^3 + 15\lambda^2 - 67\lambda + 92 = (15 - \lambda)(\lambda^2 + 24) = 0.$$

This equation has one real solution $\lambda = 15$ and two complex solutions $\lambda = \sqrt{24}i$ and $\lambda = -\sqrt{24}i$. In this example, $\lambda = 15$ is greater than the modulus of $\pm\sqrt{24}i = |\sqrt{24}| = \sqrt{24}$. That means, $\lambda = 15$ is the spectral eigenvalue (the eigenvalue with largest magnetite, or largest norm). Also this is called the spectral radius of the matrix. This spectral eigenvalue issue is true for all magic square matrices and here we will not prove this theorem in general because we are in the fun part of mathematics and in the recreational mathematics area we often do not engage in rigorous proofs. Of course, we can get $\lambda = 15$ by adding all three rows of the determinant and replacing them in the first row. Obviously all three entries along the first row will be $(15-\lambda)$ and we can factorize it and obtain $\lambda = 15$ as a solution of the characteristic polynomial.

We finish this study with a note on a very famous magic square. Let's consider the following famous Yang Hui D$^{\ddot{u}}$rer Magic Square (for a note on the name of this magic square, see [1], page 75). This magic square has four eigenvalues which are λ=34, 8, 8 and 0. Again here, the eigenvalue λ=34 is equal to the magic sum of the magic square which is S=34 and it is the largest eigenvalue between four eigenvalues. For two reasons, the inverse of this matrix does not exist. The first reason is that the determinant of this matrix is equal to zero and so it is a singular matrix with no inverse, secondly, it has an eigenvalue equal to λ=0. So, like all aspects of life, nothing is perfect. Although this magic square has many interesting properties (see [1]) it is a singular matrix with no inverse.

Yang Hui D^{u}rer Magic Square
with magic sum S=34

22

References

1. Hossein Behforooz: *Behforooz Calendarical Magic Squares*, Topics in Recreational Mathematics, vol. 3, 2015.

2. Ronald Lancaster: *Magic Squares and Matrices,* Mathematics Teachers, vol. 72 (1979), 30-32.

3. Clifford Pickover: *The Zen of Magic Squares, Circles, and Stars,* Princeton University Press, Princeton, NJ, 2002.

4. David Rose: *Magic Squares and Matrices,* the Mathematical Gazette, vol. 57 (1973), 36-39.

The Importance of Winning Draw Controls in Women's Lacrosse

Alexandra L. DeMarco

Zoe M. Loveman

Paul M. Sommers

Department of Economics

Middlebury College

Middlebury, Vermont 05753

psommers@middlebury.edu

Abstract

The authors examine the individual box scores of all men's and women's lacrosse games in 2013, 2014 and 2015 for the eleven schools in the New England Small College Athletic Conference to assess the importance of winning face offs to winning games. In men's lacrosse, there is no relationship between face offs won and games won. In women's lacrosse, however, there is a strong direct relationship between draw controls (as face offs are called in women's lacrosse) and games won. The authors then use regression analysis to find (for each of the eleven schools) the minimum percentage of draw controls won needed to win a game.

"Win the draw, win the game."

— Lauren Read [1]

Face offs (also called "draw controls" in women's lacrosse) occur at the beginning of each period (two for women, four for men) and after each goal. Face offs occur at the center of the field and determine who will receive possession of the ball.[1] A face off won is awarded to the team of the player who gains possession of the ball directly following the face off. Smith, Banker and Sommers [2] showed that ground balls are critical to success in men's lacrosse. But, how important are face offs to men's and women's lacrosse? In this paper, to gauge the importance of face offs, we examine the individual box scores of all 542 men's and 568 women's lacrosse games for all eleven Division III colleges in the New England Small College Athletic Conference (hereafter NESCAC) between 2013 and 2015.[2] Trinity won the women's NESCAC title in 2013, 2014 and 2015. All three years, Trinity was the runner-up in the NCAA Division III championship. In men's lacrosse, Tufts won the NESCAC title in 2013, 2014 and 2015. Tufts won the national title in 2014 and again in 2015.

To test the null hypothesis that face offs won is not related to games won, we use a chi-square test. All lacrosse games are either won or lost; there are no ties, with as many "sudden death" overtime periods as are necessary to break ties at the end of regulation. Although we report the outcome of games where the number of face offs for and against each team was the same, the chi-square test statistic was calculated excluding cells where the number of face offs for and against each team was the same. All games for each NESCAC school are divided into four groups as shown, for example, by one NESCAC school, Middlebury College, in Table 1.

In 18 [9] men's lacrosse games, Middlebury College recorded more [fewer] face offs than her opponent (that is, "face offs for" exceeded [was less than] "face offs against") *and* Middlebury College won [lost] the game. The calculated chi-square (χ^2) statistic is 0.536. The probability that the chi-square test statistic will be as large as this (or larger) is 0.464. This *p*-value is too high to reject the null hypothesis at the customary 5% level. That is, at this level χ^2 fails to establish any dependence between winning face offs and winning games. Middlebury College won 69 percent [= 18/(18 + 8) × 100] of all games when she won more face offs than her opponent; Middlebury still won 59 percent [= 13/(13 + 9) × 100] of all games when she did not win more face offs than her opponent. The difference between 69 percent and 59 percent is not statistically discernible.

The results for all eleven NESCAC schools are reported in Table 2. In every case but one (Wesleyan), winning more face offs than one's opponent is *not* critical to success. The results for all eleven NESCAC schools in women's lacrosse are shown in Table 3.

In all but two cases (Trinity and Williams), we can reject the null hypothesis of independence. That is, when a team won more (fewer) face offs than her opponent, more often than not that team won (lost) the game. Diagonal elements of the contingency table are disproportionately large. In women's lacrosse: win the draw, win the game.

Why are face-offs more critical to success in the women's game than in the men's? In women's lacrosse, most aggressive stick checking and body contact are illegal. Women can only stick check (to knock the ball out of their opponent's stick) if the check is stick to stick, the player's stick is below shoulder level, and no contact is made to the opposing player's body. In the men's game, more body and stick contact are allowed that would otherwise be called a foul in the women's game. Male lacrosse players therefore do not rely as heavily on the face off for possession because they can more easily check their opponent and knock the ball loose. That is, once a woman wins the draw, she is more likely to keep possession of and thus control the ball than her male counterpart. More possessions, in turn, lead to more scoring opportunities and thus a better chance of winning the game. In the words of Kelly Amonte Hiller [3, p. 92], the head women's lacrosse coach at Northwestern University:

Draw controls are the key to the game of lacrosse. Possession is the most defining factor in a game because a team cannot score without the ball, and winning draw controls is the best way to get control of the ball.

Table 4 shows the results of regressing (for each game in women's lacrosse at each NESCAC school over the three-year period, 2013-2015) the margin of victory (that is, "goals for" minus "goals against") against the percentage of draw controls (or face offs) won. How well the regression line fits the scatter of points (as measured by the R^2 or the coefficient of determination) is best for Colby. In every case, there is evidence of a strong direct relationship between the team's margin of victory and the percentage of draw controls won. One can use the regression results to find the percentage of draw controls won above which the winning margin is greater than or equal to "1" goal (as reported in the last column of Table 4). Trinity, the NESCAC champion all three years and hence the largest number (n) of games played including post-season matches, has the smallest minimum threshold (13.9 percent). That is, if Trinity wins at least 13.9 percent of the draw controls in any given game, it is likely to win that game. And, the two NESCAC schools with the worst winning records over the three years (namely, Connecticut College, .311 overall winning percentage and Wesleyan, .333) and hence the smallest number of games played (because they did not qualify for the postseason), have the largest minimum thresholds needed to win a game (63.1 percent for Connecticut College and 60.9 percent for Wesleyan).

Concluding Remarks

An analysis of all men's and women's lacrosse games for each of the eleven Division III NESCAC schools between the years 2013 and 2015 shows (with one exception) no relationship between face offs won in men's lacrosse and games won. In women's lacrosse, over the same time period, all NESCAC schools (but two) show that face offs or draw controls (as they are referred to in women's lacrosse) have a significant impact on the outcome of the game. That is, draw controls win games in women's lacrosse, but not in men's lacrosse.

Table 1
Contingency Table Relating Face Offs Won and Games Won
in Men's Lacrosse,
Middlebury College, 2013 through 2015

	Won Face Offs?	
	Yes	No
Won Game?		
Yes	18	13
No	8	9

Table 2
Summary of Chi-Square Tests,
Men's Lacrosse in NESCAC,
2013, 2014, and 2015

School	Won Game, Won FOs	Won Game, Tie FOs	Won Game, Lost FOs	Lost Game, Won FOs	Lost Game, Tie FOs	Lost Game, Lost FOs	χ^2	p-value
Amherst	17	4	17	7	0	11	0.585[a]	.444
Bates	14	1	7	11	2	9	0.586	.444
Bowdoin	6	1	13	10	0	16	0.227	.634
Colby	17	0	4	15	2	7	0.920	.337
Connecticut	21	1	3	14	1	7	2.813	*.094*
Hamilton	10	1	11	10	0	13	0.076	.783
Middlebury	18	3	13	8	0	9	0.536	.464
Trinity	4	3	6	11	1	20	0.067	.797
Tufts	36	1	20	4	0	6	2.096	.148
Wesleyan	20	1	13	4	2	11	4.752	**.029**
Williams	11	2	11	9	0	12	0.220	.639

[a]All chi-square tests are based on 2 × 2 contingency tables, excluding the two categories "Won Game, Tie FOs" and "Lost Game, Tie FOs." Otherwise, two of the six cells would have expected frequencies of less than 5 and, under these circumstances, a chi-square analysis may not be appropriate.

Table 3. Summary of Chi-Square Tests, Women's Lacrosse in NESCAC, 2013, 2014, and 2015

School	Won Game, Won FOs	Won Game, Tie FOs	Won Game, Lost FOs	Lost Game, Won FOs	Lost Game, Tie FOs	Lost Game, Lost FOs	χ^2	p-value
Amherst	27	3	6	8	2	6	3.148[a]	*.076*
Bates	16	1	5	8	1	15	7.591	**.006**
Bowdoin	30	2	7	5	2	8	8.321	**.004**
Colby	26	0	7	6	0	12	10.295	**.001**
Connecticut	7	2	5	5	3	23	6.553	**.010**
Hamilton	16	1	10	3	3	16	9.418	**.002**
Middlebury	37	3	6	8	0	5	3.799	*.051*
Trinity	37	6	18	3	0	3	0.715	.398
Tufts	21	2	6	6	2	13	9.818	**.002**
Wesleyan	10	1	4	6	4	20	8.865	**.003**
Williams	12	3	14	5	0	15	2.171	.141

[a]See footnote *a* in Table 2.

Table 4
Summary of Regression Results,
Women's Lacrosse in NESCAC,
2013, 2014, and 2015

Dependent variable: Winning margin (goals for – goals against)

School	Constant	Percentage of draw controls won	n	R^2	Percentage of draw controls won, winning margin ≥ 1
Amherst	-5.555 (-1.64)[a]	.155 (2.69)***	52	.126	42.4
Bates	-15.183 (-3.89)***	.291 (4.17)***	46	.283	55.5
Bowdoin	-13.043 (-4.04)***	.287 (5.32)***	54	.353	48.9
Colby	-11.389 (-3.48)***	.282 (4.73)***	52	.309	43.9
Connecticut	-10.061 (-3.88)***	.175 (2.97)***	44	.174	63.1
Hamilton	-4.801 (-1.75)*	.110 (1.92)*	49	.073	52.7
Middlebury	-6.217 (-1.60)	.197 (3.02)***	59	.138	36.6
Trinity	-0.523 (-0.21)	.110 (2.48)**	67	.087	13.9
Tufts	-12.405 (-3.87)***	.255 (4.27)***	50	.275	52.6
Wesleyan	-10.869 (-3.17)***	.195 (2.79)***	45	.153	60.9
Williams	-7.611 (-2.37)**	.181 (2.73)***	49	.137	47.5

[a]Numbers in parentheses are *t*-values (*$p \leq .10$, **$p \leq .05$, ***$p \leq .01$)

References

1. L. Read, "Win the draw, win the game." *Burlington Free Press*, May 31, 2015, p. 17A.

2. P. R. Smith, R. K. Banker, and P. M. Sommers, "Here's the scoop: ground balls win lacrosse games." *Topics in Recreational Mathematics* (ed. by Charles Ashbacher), Vol. 2, 2015, pp. 17-27.

3. K. A. Hiller, A. Gersuk (contributor), and A. Elliott (contributor), *Winning Women's Lacrosse*, Human Kinetics, 2009.

Radical Axis of Lemoine Circles

Ion Patrascu

Professor "Fraţii Buzeşti" National College,

Craiova, Romania

Florentin Smarandache

Professor, New Mexico University, USA

Abstract

In this short paper, a theorem stating that the radical axis of of the Lemoine circles of a triangle is perpendicular to a line raised on the symmedian is proven.

In this article, the emphasis is on the radical axis of the Lemoine Circles. We open with some definitions.

Definition: In a triangle, **a symmedian** is a line constructed by first drawing a line from a vertex to the midpoint of the opposite side and then reflecting that line across the line that bisects the angle. You will then have three segments emanating from a vertex with the angle bisector in the middle and the symmedian on the side opposite the segment connected to the midpoint of the opposite side. There are of course three symmedians in a triangle.

Definition: The **symmedian center** of a triangle is the point where the three symmedian segments mutually intersect.

Definition: A **Lemoine parallel** is a line through the symmedian point of a triangle that is parallel to a side of the triangle.

Definition: A **Lemoine circle** is the circle determined by the points where the Lemoine parallels intersect the sides of the triangle.

We open by reminding the reader of the statements of two theorems.

Theorem 1: The parallels taken through the symmedian center K of a triangle to the sides of the triangle determine on them six concyclic points (The First Lemoine Circle).

Theorem 2: The antiparallels taken through the symmedian center of a triangle to the sides of a triangle determine on them six concyclic points (The Second Lemoine Circle).

Remark 1 : If ABC is a scalene triangle and K is its symmedian center, then L, the center of the First Lemoine Circle, is the middle of the segment [OK], where O is the center of the circumscribed circle, and the center of the Second Lemoine Circle is K. It follows that the radical axis of Lemoine circles is perpendicular on the line of the centers LK, therefore on the line OK.

Proposition 1 : The radical axis of Lemoine Circles is perpendicular on the line OK raised in the symmedian center K.

Proof : The proof references figure 1. Let $A_1 A_2$ be the antiparallel to BC taken through K, then KA_1 is the radius R_{L_2} of the Second Lemoine Circle; we have:

$$R_{L_2} = \frac{abc}{a^2 + b^2 + c^2} .$$

34

Figure 1

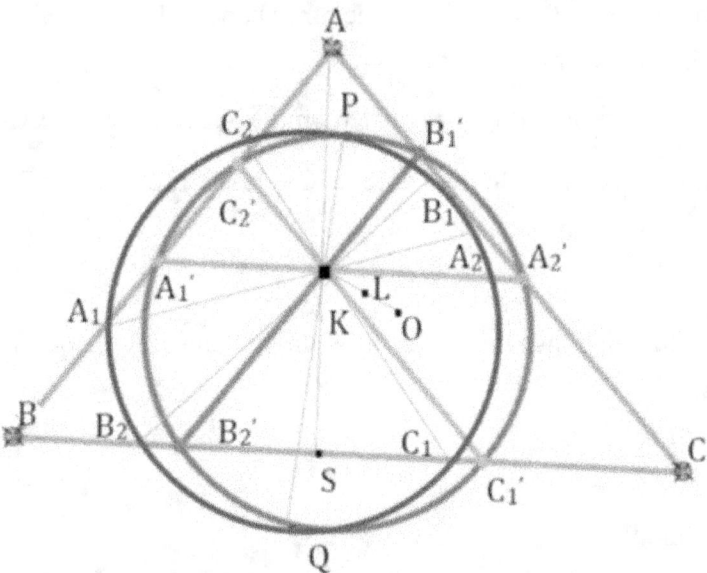

Let $A_1' A_2'$ be the Lemoine parallel taken to BC; we evaluate the power of K towards the First Lemoine Circle. We have:

$$\overrightarrow{KA_1'} \cdot \overrightarrow{KA_2'} = LK^2 - R_{L_1}^2. \tag{1}$$

Let S be the simmedian leg from A; it follows that:

$$\frac{KA_1'}{BS} = \frac{AK}{AS} - \frac{KA_2'}{SC}.$$

We obtain:

$$KA_1' = BS \cdot \frac{AK}{AS} \quad \text{and} \quad KA_2' = SC \cdot \frac{AK}{AS},$$

but $\frac{BS}{SC} = \frac{c^2}{b^2}$ and $\frac{AK}{AS} = \frac{b^2+c^2}{a^2+b^2+c^2}$.

Therefore:

$$\overrightarrow{KA_1'} \cdot \overrightarrow{KA_2'} = -BS \cdot SC \cdot \left(\frac{AK}{AS}\right)^2 = \frac{-a^2b^2c^2}{(b^2+c^2)^2} \cdot \frac{(b^2+c^2)^2}{(a^2+b^2+c^2)^2} = -R_{L_2}^2 . \tag{2}$$

We draw the perpendicular in K on the line LK and denote by P and Q its intersection to the First Lemoine Circle; we have $\overrightarrow{KP} \cdot \overrightarrow{KQ} = -R_{L_2}^2$; by the other hand, $KP = KQ$ (PQ is a chord which is perpendicular to the diameter passing through K).

It follows that $KP = KQ = R_{L_2}$, so P and Q are situated on the Second Lemoine Circle.

Because PQ is a chord which is common to the Lemoine Circles, it follows that PQ is the radical axis.

Comment 1: After equalizing relations (1) and (2) or by the Pythagorean theorem in the triangle PKL, we can calculate R_{L_1}. It is known that:

$$OK^2 = R^2 - \frac{3a^2b^2c^2}{(a^2+b^2+c^2)^2},$$

and since $LK = \frac{1}{2}OK$, we find that:

$$R_{L_1}^2 = \frac{1}{4} \cdot \left[R^2 + \frac{a^2b^2c^2}{(a^2+b^2+c^2)^2}\right].$$

Remark 2 : The proven *Proposition* regarding the radical axis of the Lemoine Circles is a particular case of the following *Proposition*, which we leave it to the reader to prove.

Proposition 2: If $\mathcal{C}(O_1, R_1)$ şi $\mathcal{C}(O_2, R_2)$ are two circles such as the power of center O_1 towards $\mathcal{C}(O_2, R_2)$ is $-R_1^2$, then the radical axis of the circles is the perpendicular in O_1 on the line of centers O_1O_2.

References

1. F. Smarandache, Ion Patrascu: *The Geometry of Homological Triangles*, Education Publisher, Ohio, USA, 2012.

2. Ion Patrascu, F. Smarandache: *Variance on Topics of Plane Geometry*, Education Publisher, Ohio, USA, 2013.

Word Hypercubes are Fun, NP-Hard, and In General Undecidable

Barry Fagin
Leemon Baird

Abstract

Word-hypercubes are a generalization of word squares for more than 2 dimensions. Given a dictionary of words of length n, and a d-dimensional hypercube partitioned into n^d spaces of dimension d, can letters be placed into all spaces so that words from the dictionary are formed when reading unidirectionally? We show that this problem is NP-complete, and also give examples of both new word squares and the word hypercube of highest dimensionality known to the authors.

Introduction

Every year, the University of Chicago organizes a scavenger hunt on campus. This is like saying Genghis Khan organized tea parties in central Europe. A UC scavenger hunt involves road trips, ridiculously hard-to-find items ("your appendix" was one), and silly stunts. "Scav" also includes plenty of brain teasers, because the University of Chicago is that kind of place.

The list of items participants must find and bring to the judges for scoring is routinely made public, so that parents and alumni can join in the madness. We call the readers' attention to item #345 from the 2012 list [3]:

> #345. A four-letter word. No, wait. A four-letter word square. No, wait! A four-letter word cube. NO, WAIT! A four-letter word tesseract. [4^1 points for a word cube, 4^2 points for a word tesseract]

A four-letter word square is a 4x4 grid filled in with four words such that all squares are symmetric about the diagonal. For example, figure 1 is a four-letter word square:

Figure 1

S	A	F	E
A	R	E	A
F	E	A	R
E	A	R	N

Teams would be awarded four points for extending this idea in the obvious way to three dimensions, sixteen points if they could stretch it out to four.

This got us thinking about higher-order word objects. How many are there? What happens to the size of the solution space as the number of dimensions increases? As the word length increases? What is involved in coding up the problem for solution by computer? This article discusses what we have learned.

Previous Work and Definitions

Numerous articles in "Word Ways" [1,4,6,8,9,10,11,12] discuss word squares and word cubes of various sizes. Kon [12] discusses the concept of *word hypersolids*, the extension of word squares and cubes into different shapes and/or higher dimensions. Such objects can be either symmetric, where the words are the same in each dimension, or asymmetric, where they are allowed to differ. We will concern ourselves only with objects of equal length in all dimensions, which we will call *word hypercubes,* or simply *hypercubes* or even *cubes* when the context is clear.

Noting as Kon does that the terminology remains somewhat ambiguous, we restrict our analysis to what he calls "single" objects: Symmetric word squares extended in the obvious way to cubes, tesseracts, and so forth. For notational purposes of this article, (w,d) will denote a single word hypercube made from words of size w in d dimensions. Thus a (4,2) hypercube is a 4x4 word square, a (5,3) hypercube is a 5x5x5 word cube, and so forth.

For example, a $d = 2$, $w = 9$ word square can be thought of as a square filled with letters, where $W_{r,c}$ is the letter on row r and column c. Then $W_{1,1}$ would be the upper-left corner, and $W_{9,9}$ the lower-right.

If $d = 4$ and $w = 9$, it becomes a 4-dimensional hypercube, with the letter $W_{1,1,1,1}$ at one corner, and $W_{9,9,9,9}$ at the opposite corner. If we leave off the last coordinate, we get a word. So $W_{4,2,7}$ is the 9-letter word made up of the letters $W_{4,7,2,1}$, $W_{4,7,2,2}$, ..., $W_{4,7,2,9}$. In a valid word hypercube, this word must be in the dictionary.

The word hypercube has words crossing each other, so there is one additional constraint: two letters must be equal if they have the same subscripts, but in a different order. So we must choose the letters such that $W_{4,7,2,1} = W_{1,2,4,7}$, and $W_{4,7,2,9} = W_{2,4,7,9}$. This would also imply that words must be equal when their subscripts are rearranged, so the word $W_{4,7,2}$ must be the same as word $W_{2,4,7}$.

Given these constraints, we could choose to fill a word hypercube by choosing only the words and letters whose subscripts are sorted in ascending order. So we would choose the letter $W_{1,2,4,7}$, but would not explicitly choose $W_{4,7,2,1}$, because the latter is determined by the former. Similarly, we would only check words against the dictionary if they have sorted subscripts. So we would check $W_{2,4,7}$, but not $W_{4,7,2}$.

We note there is another way to write a word hypercube that may be easier to visualize in some cases. For example, when $w = 3$ and $d = 5$, the standard word hypercube is a 3 x 3 x 3 x 3 x 3 hypercube containing 243 letters, and 81 words, with each word consisting of 3 letters in a straight line. But many of those letters and words are constrained to be equal, so there are only 21 letters that can be chosen independently, and only 15 words that must be checked in the dictionary. If those 21 letters are written in the format seen in figure 2, then each 3-letter word will be L-shaped rather than a straight line. In this format, a word is the letter in a cell, followed by the letter in the cell immediately above it, followed by the cell immediately to the right. So the word starting at the lower-left consists of these three letters in this order: $W_{1,1,1,1,1}$, $W_{1,1,1,1,2}$, $W_{1,1,1,1,3}$. There are 15 such L-shaped words in this diagram. All 15 of those words are contained in the dictionary, if and only if this diagram represents a valid 3x3x3x3x3 word hypercube.

Figure 2

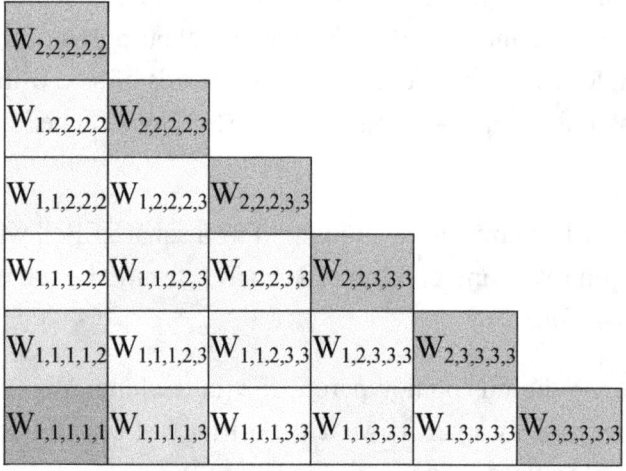

The validity of this format is clear. The letters that can be chosen independently are those that have their subscripts in sorted order. Since each subscript can be only a 1, 2, or 3 (for w = 3), we can summarize the list of subscripts by giving only the number of 2's, the number of 3's, and the length of the list. In the above diagram, the number of 2's increases vertically, and the number of 3's increases horizontally.

This format also makes it easier to visualize what happens when the dimensionality d is changed. If d is decremented by one, then the top diagonal (purple) is deleted, and the first subscript (1) is deleted from all the remaining letters. This can be repeated. Each time d is decremented, one more diagonal is deleted, and another 1 subscript is deleted. Conversely, to increment d, first prepend a 1 to every list of subscripts, then add a new diagonal above the top one, with the subscripts following the same pattern as seen in the purple: all 2's at the top, with more 3's shifting into each cell as you move down, reaching all 3's at the bottom.

This pattern also continues for larger values of w. When w = 3, it is a 2 - dimensional triangle, as shown above. In general, it is a (w-1)-dimensional hyperpyramid. For w = 3, a word is the letter in a cell, followed by the cells immediately adjacent to it in the positive horizontal and vertical directions. In general, a word is the letter in a cell followed by the cells immediately adjacent to it in the positive direction parallel to each of the (w-1) axes.

For a (w,d) cube, it is not difficult to show that the number of words $N(w,d)$ required is given by

$$N(w,d) = \begin{pmatrix} w + d - 2 \\ d - 1 \end{pmatrix}$$

where parentheses denote the binomial coefficient. Similarly, the number of letters $L(w,d)$ is given by

$$L(w,d) = N(w,d+1)$$

Word Hypercubes Are NP-Hard

How hard is it to find a word hypercube of a given word size and dimensionality?

Unfortunately for those who hope to discover new ones, it is NP-hard.

For this section, we assume the reader has a general familiarity with computational complexity theory. Readers who do not may skip this section, noting only that finding word hypercubes using a given word list is a member of a class of problems widely believed to require exponentially increasing time to solve as the word list gets larger.

Given a problem (w,d,D,A) where:

- w = the word length (a positive integer)
- d = dimensionality (a positive integer)
- A = an alphabet (set of symbols)
- D = a dictionary (a set of strings over A of length w)

The Word Square Problem is the question: Does there exist a word square of size w for dictionary D? This is the $(w,2,D,A)$ problem. The Word Hypercube Problem is the general (w,d,D,A) problem in d dimensions. We believe the complexity of this problem was previously unknown, even for the $d = 2$ Word Square Problem. Gary and Johnson [7] list the Crossword Puzzle Problem as NP-complete, but that problem lacks the symmetry requirements of the Word Square Problem, so it does not prove the Word Square Problem to be NP-complete. It is also unrelated to the higher-dimensional generalization.

Theorem: The Word Hypercube Problem is NP-hard.

Proof: The proof is by reduction from 3SAT [7]. We will first construct a dictionary that is guaranteed to have exactly one word-hypercube solution. Then we will modify that dictionary to embed an arbitrary 3SAT problem.

Consider the word hypercube problem with w = d = 4 and a dictionary of four words D = {ABBB, BCCC, CDDD, DEEE}. Figure 3 shows one solution, with the dictionary on the left, and the solution on the right, with letters shaded for clarity:

Figure 3

```
A B B B    B C C C    B C C C    B C C C
B C C C    C D D D    C D D D    C D D D
B C C C    C D D D    C D D D    C D D D
B C C C    C D D D    C D D D    C D D D

B C C C    C D D D    C D D D    C D D D
C D D D    D E E E    D E E E    D E E E
C D D D    D E E E    D E E E    D E E E
C D D D    D E E E    D E E E    D E E E

B C C C    C D D D    C D D D    C D D D
C D D D    D E E E    D E E E    D E E E
C D D D    D E E E    D E E E    D E E E
C D D D    D E E E    D E E E    D E E E

B C C C    C D D D    C D D D    C D D D
C D D D    D E E E    D E E E    D E E E
C D D D    D E E E    D E E E    D E E E
C D D D    D E E E    D E E E    D E E E
```

```
A B B B

B C C C

C D D D

D E E E
```

In the solution, the hypercube element at location (1,1,1,1) is the A in the upper-left corner of the upper-left square, and the element at location (4, 4, 4, 4) is the"E" in the lower-right corner of the lower-right square.

The dictionary word ABBB appears in the hypercube four times: twice in the upper-left square, once going down through the layers shown by the top row of squares, and once going down through the layers shown by the left column of squares. This word could not have been placed anywhere else. If it had been placed at another location, then the "A" would have appeared somewhere other than location (1,1,1,1), and so would have had to intersect some word at a position other than its first letter. But no word in the dictionary has an "A" in any position other than the first position. So this word could not have appeared anywhere else.

In fact, the solution shown is the only possible solution for this dictionary. If the word ABBB is part of the solution, then it will have to appear in the position shown. And then BCCC will have to appear in the position shown, because its leading "B" must intersect with a word that has a "B" in a position other than first, and ABBB is the only word that has that. And then CDDD and DEEE must appear as shown, by the same argument. So if ABBB is used at all, then the only solution will be the one shown.

On the other hand, what if ABBB is not used? Then suppose BCCC is used. Since no other word contains a "B", the BCCC will have to appear where ABBB is shown

42

in the diagram. That forces CDDD to move to where BCCC is shown, and DEEE to move to where CDDD is shown. But then there is no word that can go in the positions where DEEE are shown. A word in that position would have to start with E. But there are no words in the dictionary starting with E. So there cannot be a solution that contains BCCC without ABBB.

By a similar argument, there is no solution with CDDD and no BCCC or ABBB and there is no solution with DEEE without the other words. Therefore, the solution shown is unique.

This generalizes in the obvious way to word lengths other than $w = 4$. The words simply expand or contract by using more or fewer repetitions of the final letter. It also generalizes to dimensions other than $d = 4$ in the obvious way. The dictionary will contain d words, with each word starting with the letter that ended the previous word, followed by repetitions of a new letter.

Now we will delete the first 3 words of the dictionary and replace them with a longer list of words. In the process, the letters B and C will be removed from the alphabet, and several subscripted versions of them will be added. The new dictionary is:

$$D = \{ AB_1B_2B_3...B_{w-1},$$

$$B_1C_0C_0C_0...C_0,$$

$$B_2C_0C_0C_0...C_0,$$

$$...$$

$$B_{w-1}C_0C_0C_0...C_0,$$

$$C_0DDD...D,$$

$$C_1DDD...D,$$

$$DEEE...E \}.$$

In high dimensions, this could be a very long list, but it will still only be the first 3 words that were changed. This new dictionary again has only a single possible solution. The first square in that solution for the $w = 4$ case now looks like what is seen in figure 4.

Figure 4

A	B_1	B_2	B_3
B_1	C_0	C_0	C_0
B_2	C_0	C_0	C_0
B_3	C_0	C_0	C_0

This is identical to before, except for the subscripts and it is unique for the same reason as before. Plus the single word starting with A forces all the B words to be in a particular order.

Now consider the word $B_2C_0C_0C_0$ in the dictionary. It is forced to be on the third row and the third column in the solution. If the word $B_2C_0C_1C_0$ were added to the dictionary, it could also be placed on the third row and column, so there would now be two valid solutions. In fact, any combination of the zero subscripts in $B_2C_0C_0C_0$ could be replaced with ones, and if the resulting words were added to the dictionary, each would give a valid solution if it were placed on the third row and column. Similar words starting with B_1 or B_3 could also be added, which might generate additional solutions, if they could be chosen such that their intersecting C letters matched.

The image of figure 5 is the same square for a larger word length w, where an arbitrary rectangle has been chosen in the yellow region above the diagonal, and it has been outlined with a solid line, with its transpose outlined with a dotted line.

Figure 5

A	B_1	B_2	B_3	B_3	B_3	B_3	B_3	B_3	B_3
B_1	C_0	C_0	C_0	C_0	C_0	C_0	C_0	C_0	C_0
B_2	C_0	C_0	C_0	C_0	C_0	C_0	C_0	C_0	C_0
B_3	C_0	C_0	C_0	C_0	C_0	C_0	C_0	C_0	C_0
B_4	C_0	C_0	C_0	C_0	C_0	C_0	C_0	C_0	C_0
B_5	C_0	C_0	C_0	C_0	C_0	C_0	C_0	C_0	C_0
B_6	C_0	C_0	C_0	C_0	C_0	C_0	C_0	C_0	C_0
B_7	C_0	C_0	C_0	C_0	C_0	C_0	C_0	C_0	C_0
B_8	C_0	C_0	C_0	C_0	C_0	C_0	C_0	C_0	C_0
B_9	C_0	C_0	C_0	C_0	C_0	C_0	C_0	C_0	C_0

Such a rectangle can be chosen arbitrarily, and then each row and column within that rectangle can be given a set of possible patterns by adding to the dictionary the words containing those patterns. Since the rectangle is above the diagonal, it will not overlap with its transpose, which simplifies the analysis.

Now it is possible to reduce 3SAT to the Word Hypercube problem of any desired dimensionality. Suppose we are given a 3SAT problem in conjunctive normal form with v variables and n clauses, and no repeated variables within a single clause. We can now generate a word hypercube problem of the desired dimensionality and with a word length long enough so that a rectangle with v rows and n columns can fit within it, as in the above figure.

Each row of the rectangle will correspond to one variable in the 3SAT problem. Each column of the rectangle will correspond to one clause in the 3SAT problem. The dictionary will contain 2 words for each row, and 7 words for each column

For each row in the rectangle, the dictionary has two words. One word has a C_0 for every position that falls within the rectangle. The other word is the same, except there is a C_1 in each column that corresponds to a 3SAT clause that contains that variable.

A 3SAT clause will contain 3 variables, each of which might be negated, so the clause will be satisfied by 7 of the 8 possible assignments of values to those variables. So for each column in the rectangle, the dictionary will contain 7 words corresponding to the 7 satisfying assignments. Each word will have a C_1 in the position corresponding to each variable that is true in that assignment, and will have a C_0 for those variables that are false, and for the variables that don't appear in the clause at all.

Given this dictionary, it will be possible to construct a valid word-hypercube solution if and only if it is possible to make a valid word-square solution in the first square layer, which includes the first word in the dictionary. It will be possible to make that word-square solution if and only if the 3SAT problem has a satisfying assignment. Thus 3SAT has been reduced to Word Hypercube, and so Word Hypercube is NP-hard.

Corollary: The Word Square Problem is NP-complete.

Proof: The proof for the general Word Hypercube Problem works for any $d > 1$, so it shows that even for $d = 2$, it is still NP-hard. It is obvious that a Word Square Problem solution can be checked in polynomial time, and so is in NP. Therefore the Word Square Problem is NP-complete.

Mapping The Space Of Word Hypercubes Using The English Open Word List (EOWL)

Although finding word hypercubes is NP-Hard, most English words are sufficiently short and computers sufficiently fast that many parts of the (w,d) solution space can be mapped for a specific dictionary. For our results, we used the English Open Word List, or EOWL [5]. This list has the advantage of being in the public domain and therefore readily obtainable. It contains approximately 130,000 words, all under ten characters, with no words that contain hyphens, diacritical marks, or apostrophes.

For the EOWL and most values of w and d , our results show whether or not a word hypercube exists, and can give an example. For smaller values of d, we know all the solutions. Because the construction of a (w,d) hypercube requires at a minimum the existence of *w* hypercubes of dimensionality d - 1, for larger values of d we can tell whether or not objects of that dimensionality exist for some dictionary by merely inspecting the solution space of the dimension below. It is in the middle where the most mystery remains.

The results shown here were obtained using the programming language Python, run on the first author's computer or on the cloud at www.picloud.com. The use of word lists other than EOWL will of course produce different results. Running the code with larger word lists or word lists from different languages is one possible area of future work.

The results below shows what is known of the EOWL word hypercube space so far. In some cases, the number of hypercubes is known exactly. In other cases (based on our self-imposed limit of no more than one week of CPU time per problem) it is estimated, and in at least one case the existence of a solution is not yet known. We elaborate on some details in table 1.

Table 1

The EOWL Word Hypercube Space

w	dictionary size	d=2	d=3	d=4	d>4
3	1,083	132,454	37,375,325	22×10^9 (est.)	?
4	4,447	4,144,949	36×10^9 (est.)	427×10^{12} (est.)	?
5	9,636	10,724,619	610×10^6 (est.)	19×10^6 (est.)	?
6	16,219	1,652,918	22	0	0
7	22,923	9,744	0	0	0
8	26,265	0	0	0	0
9	25,626	0	0	0	0
10	22,679	0	0	0	0

For w = 3 and w = 4, the time requirements for d = 4 and higher exceed our patience thresholds to enumerate the search space. However, the inclusion in the EOWL of word cycles of length w like { EAT,ATE,TEA} and { ANAN, NANA, ANAN, NANA} means that hypercubes of arbitrary dimensionality can be constructed for w = 3 and w = 4. Thus if we allow solutions that reuse words, we would expect these search spaces to be quite large. Intermittent computational runs with samples of the solution space suggest that for w = {3,4}, the number of solutions does indeed grow rapidly for d = 4 and higher.

For w = 5, the computing requirements exceed our resources and patience to enumerate the search space beyond d = 2. Partial runs with sampling suggest that the number of solutions for (5,3) is approximately thirty times that for (5,2). Continued partial computation for (5,4) suggests that the number of solutions falls off by about two orders of magnitude. We do not yet know if a (5,5) hypercube can be formed from words in the EOWL.[1]

We will say more about hypercubes with w ≤ 5, d ≥ 3 shortly.

Although it may seem that one must exhaustively search the entire space to determine whether or not a (w,d) hypercube exists, that is not necessarily true. If the space of all (w,d - 1) and (w,d - 2) cubes is known, and it is sufficiently small, that space can be examined to determine if it contains cubes that can be assembled into a (w,d) solution. This is done by attempting to place w (w,d - 2) cubes in a w x w grid G such that a) G(i,j) = G(j,i) for all i, j between 1 and w, and b) all rows of G are (w,d - 1) cubes. (Note that the first requirement means that the columns of G will be those same cubes). For larger values of d with completely enumerated (w,d - 1) and (w,d - 2) solution spaces, this is more efficient than exhaustive search of the (w,d) space.

For example, a 5 - cube requires its first and second 5-square slices to share a word at their second and first positions, respectively. If all the 5-squares are known, and no two have this property, then no 5-cubes exist. Clearly this is not the case for (5,3) cubes, but the technique generalizes to higher values of w and d where it can be more efficient. And of course in some cases the solution set for (w,d - 1) is so small that the nonexistence of (w,d) cubes can be concluded by inspection.

Note that our results show no word squares for w > 7. Such objects are in fact known, but are drawn from different word lists [1,14]. The closest example that can be obtained from the EOWL is below, in which all but one letter was successfully placed:

---

[1] Borgmann [2] claims the longest English word cycle is {ESTER, STERE, TERES, EREST, RESTE}. Any list containing these words can therefore be used to construct a 5-cube of arbitrary dimensionality. Of these five words, only ESTER and STERE are in the EOWL. The EOWL does contain the tantalizing {PESTO, ESTOP, STOPE, TOPES] partial cycle, but no word list known to the authors contains OPEST.

```
D I S S U A D E
I M P I N G E S
S P A N * E S S
S I N F O N I A
U N * O S T L Y
A G E N T I V E
D E S I L V E R
E S S A Y E R S
```

The closest approximation to an EOWL 9 - square places all but 5 letters. For the 10 - square, the best attempts leave 12 letters unplaced.

Restricting the Solution Space to Diagonal Hypercubes

Because the solution space for $w = 3$ and $w = 4$ is so large, we consider making the problem a little more interesting by requiring one or both diagonals to be words. For higher dimensional objects, this requirement extends to all its lower dimensional slices. Unfortunately, this requirement is too stringent for $d \geq 3$, as no hypercubes meet it. For $d = 2$, the solution space appears in table 2.

If we relax the requirement so that only the main diagonal must be a word (from upper left to lower right in the square case with the obvious generality for $d > 2$), the space becomes populated for three dimensions, but not beyond, as summarized in table 3.

Table 2

EOWL Dual Diagonal Hypercubes

w	d=2
3	2083
4	7717
5	862
6	16
≥ 7	0

Table 3

EOWL Main Diagonal Hypercubes

w	d = 2	d = 3	d ≥ 4
3	9,521	4,897	0
4	140,377	2,039	0
5	67,198	0	0
6	1,506	0	0
7	2	0	0

Some Interesting Word Hypercubes

The EOWL solution space contains 22 (6,3) word hypercubes, grouped into two *families*. A family is a set of two or more word cubes closed under a set of single-letter substitutions. These families are now listed.

Family of 2

PRESTO   RUSHES   ESCORT   SHOVEL
    TERETE   OSTLER

RUSHES   UNTAME   STALER   HALITE   EMETIN
    SERENE

ESCORT   STALER   CAGILY   OLIVES
    RELENT   TRYSTS

SHOVEL   HALITE   OLIVES   VIVERS
    ETERNE   LESSEE

TERETE   EMETIN   RELENT   ETERNE
    TINNER   ENTERA

OSTLER   SERENE   TRYSTS   LESSEE
    ENTERA   RESEA**1**

1 = {L,T}

Family of 20

1ACHES	ACHENE	CHARRS	HEREAT	ENRACE	SESTET
ACHENE	CYANIN	HAREEM	ENERVE	NIEVES	ENMESH
CHARRS	HAREEM	ARGALA	REARER	RELENT	SMARTY
HEREAT	ENERVE	REARER	ERRORS	AVERSE	TERSER
ENRACE	NIEVES	RELENT	AVERSE	CENS**2**R	ESTERS
SESTET	ENMESH	SMARTY	TERSER	ESTERS	THYRS**3**

1 = {C,L,N,R,T}

2 = {E,O}

3 = {E,I}

Previously, only one 6-cube from any word list was known to the authors [10], so we believe these are new.

The EOWL word list yields a family of two 7-squares where the main diagonal is also a word:

```
1   R   O   P   P   E   R
R   E   C   L   I   N   E
O   C   R   E   A   T   E
P   L   E   A   S   E   D
P   I   A   S   T   R   E
E   N   T   E   R   E   R
R   E   E   D   E   R   S
```

1 = {C,D}

8-squares with main diagonals as words are known [14], but require the use of a different word list.

An Unusual (3,4) Hypercube

If we expand our list of 3-letter words slightly, we can construct a (3,4) word hypercube in which all main diagonals are words as well. The (3,3) slices of a such an object are as follows:

Cube 1:

PRO	RHO	OOT
RHO	HEP	OPE
OOT*	OPE	TET

                (cube diagonal is PET)

*The diagonal of this square is PHT, which appears in some word lists

Cube 2:

RHO	HEP	OPE
HEP	EEL	PLY
OPE	PLY	EYE

                (cube diagonal is REE)

Cube 3:

OOT	OPE	TET
OPE	PLY	EYE
TET	EYE	TEE

                (cube diagonal is OLE)

This object uses the words OOT, TET, PHT and OLE, which do not appear in the EOWL but do appear in other sources. The main diagonal of the entire (3,4) cube is the upper left back corner of cube 1 = 'P', the center letter of cube 2 = 'E', and the lower right front corner of cube 3 = 'E'. While regrettably scatological, the word is nonetheless included the EOWL.

Restricting the Solution Space to Hypercubes With Unique Word Placement

To reduce the solution space for shorter word lengths and make the problem more interesting, we may add the restriction that a word can only be placed once. This ensures that for any given word list, there is sufficiently high dimensionality beyond which no cubes exist. Our results and estimates for small d are shown in table 4.

Table 4

EOWL Single Word Placement Hypercubes

w	d = 2	d = 3	d = 4	d ≥ 5
3	132,187	35,774,512	2×10^9 (est.)	?
4	3,989,868	34×10^9 (est.)	121×10^{12} (est.)	?
5	9,509,258	571×10^6 (est.)	14×10^6 (est.)	?
6	1,308,120	22	0	0
7	9,744	0	0	0

An Undecidable Problem

Returning to the problem as originally stated (in which the reuse of words is allowed), given a dictionary of w-letter words, it is natural to ask what dimensionalities of word hypercubes are possible. Clearly, a $d = 1$ solution can be found: it's just a single word from the dictionary. If the dictionary contains a word that is just repetitions of a single letter, then solutions will exist for all positive d. If the dictionary looks like the first one that was constructed in the NP-completeness proof, then solutions will exist for all d up to a certain threshold, and for no d above that. So it is natural to ask the question: Does a given dictionary allow solutions for all positive d, or only for d up to some threshold value? That question isn't merely NP-hard, it's actually undecidable. No computer program can be written that will correctly answer this question for all dictionaries, even if that program is given unlimited time and memory.

Theorem: The question of whether a given dictionary with words of length w allows for word hypercube solutions for all positive d is undecidable.

Proof: The proof is by reduction from the Wang tiling problem [13].

A Wang tiling problem consists of a given, finite set of square tiles, with each edge of each tile colored a single color. Different edges can have different colors or the same color, but a single edge is assigned only one color. The problem is to tile the infinite plane with duplicates of these tiles, without rotating them, such that edges in contact are the same color. The question of whether this is possible for a given set of tiles is undecidable. The problem remains undecidable if the question is whether the infinite quarter-plane can be tiled (i.e., the first quadrant in a Cartesian system).

Given a particular set of n Wang tiles using m colors, we will construct a dictionary for the word hypercube problem, where the alphabet has $4n + m + 3$ letters, each word is $w = 3$ letters long, the dictionary contains $8n + 1$ words. We will then show that this dictionary has solutions for all positive d, if and only if the given set of Wang tiles can tile the quarter plane.

For the word hypercube problem, the alphabet will contain one letter for each color used by the Wang tiles, plus the three letters {X,Y,Z}, plus the 4n letters {A_i, B_i, C_i, D_i} for $1 \leq i \leq n$. The dictionary will contain the single, 3-letter word "XYZ", plus the 8n words given by the following, for all $1 \leq i \leq n$:

$$W_i \quad X \quad A_i$$

$$A_i \quad Z \quad B_i$$

$$B_i \quad N_i \quad E_i$$

$$Y \quad W_i \quad C_i$$

$$C_i \quad A_i \quad D_i$$

$$D_i \quad B_i \quad Y$$

$$Z_i \quad C_i \quad S_i$$

$$S_i \quad D_i \quad X$$

This defines a group of 8 words associated with each value of i. Each group together corresponds to the ith Wang tile. The four variables $\{N_i, S_i, E_i, W_i\}$, represent the four colors that are on the North, South, East, and West side of the ith Wang tile, respectively.

For the word hypercube problem for a given dimensionality d, a solution exists if and only if it is possible to fill in a triangular table of d + 1 rows and columns, similar to the one in the previous section, such that every L-shape of 3 cells (a cell, the one above, the one to the right, in that order) corresponds to a word in the dictionary. For the problem of deciding whether there exist hypercubes for all positive d, this corresponds to expanding the triangle in the diagram forever. In other words, the problem is to choose letters for every cell in the infinite quarter plane (i.e., the first quadrant) such that every such L-shape contains a valid word from the dictionary.

Note that this problem cannot be solved using only the letters $\{X, Y, Z\}$, because there is only one word that uses only those letters, and it uses Y and Z but starts with X. So it will be necessary to use at least one word from the tile groups.

If at least one word from the tile group i is to be used, then notice the effect of the letters $\{A_i, B_i, C_i, D_i\}$, given how they are distributed among the 8 words in that group. These 4 letters act as "glue". Every word in the group includes at least one of those 4 glue letters. If any word from that group is used, then at least one glue letter will appear in the table, and that will require that at least one word be used that starts with that letter, ends with that letter, and has that glue letter in the middle. The same is true for any glue letters that appear in those words, and so on. It quickly becomes apparent that if even a single word in a tile group is used, then all 8 words in that group must be used, and they must be arranged exactly like what is seen in figure 6.

Figure 6

X	Z	N_i	X
W_i	A_i	B_i	E_i
Y	C_i	D_i	Y
X	Z	S_i	X

The 4 glue letters appear in the center, and ensure that all the letters in this diagram must be as shown. The X in the lower-left corner is forced by the fact that there is a Y above it and a Z to its right, and the only dictionary word ending in "YZ" is the word "XYZ". The X in the upper-right corner is forced by process of elimination, given the reasoning below. The rest of the letters are forced by the words in the tile group.

Since the above 4 x 4 pattern must appear anywhere a tiling group word is used from the dictionary, then the only way to tile the quadrant will be to tile it using copies of this pattern, where the rightmost column of one pattern overlaps the leftmost column of the pattern to its right, and the top row of one pattern overlaps the bottom row of the pattern above it. For example, a 7 x 7 region might be filled in the manner seen in figure 7.

In this example, the upper-left region is filled with words from tile group 1, the upper right is 2, and the bottom is 3 and 4. Since the patterns are overlapping, this will only be a legal configuration if it happens to be the case that E1 = W2, and S1 = N3, and S2 = N4, and E3 = W4. Similarly, larger regions can be filled with letters only by laying them out in the same way that tiles would have covered a larger region in the Wang tiling problem.

Figure 7

X	Z	N_1	X	Z	N_2	X
W_1	A_1	B_1	$E_1{=}W_2$	A_2	B_2	E_2
Y	C_1	D_1	Y	C_2	D_2	Y
X	Z	$S_1{=}N_3$	X	Z	$S_2{=}N_4$	X
W_3	A_3	B_3	$E_3{=}W_4$	A_4	B_4	E_4
Y	C_3	D_3	Y	C_4	D_4	Y
X	Z	S_3	X	Z	S_4	X

In other words, a region (e.g., the infinite quarter plane quadrant) can be filled with letters that satisfy the constraints of the word hypercube problem, if and only if they are chosen to correspond to tiles that satisfy the constraints of the Wang tiling problem. Thus the two problems are equivalent, and if one is undecidable, then so is the other. The Wang tiling problem is undecidable, so the word hypercube problem is undecidable, and the proof is complete.

Corollary: The word hypercube problem remains undecidable, even if the word length w is constrained to be $w = 3$.

Proof: The above proof only used $w = 3$, so this is sufficient.

Theorem: The word hypercube problem is decidable for $w < 3$.

Proof: Since w is a positive integer, this can only be $w = 1$ or $w = 2$. For $w = 1$, a single dictionary word can solve the problem for all d by filling the quadrant with its single letter, so the problem is trivially decidable (since it's always true). For $w = 2$, the dictionary words are all 2-letter words, and the problem can be found by discovering whether there exists a sequence of n words from the dictionary that form a cycle of the form $\{L_1L_2,\ L_2L_3,\ L_3L_4,\ \dots,\ L_{n-1}L_n,\ L_nL_1\}$. There are solutions for all d if and only if such a cycle exists. And its existence can be found in polynomial time by constructing a finite directed graph and looking for directed cycles. Therefore, the $w = 2$ case is decidable, too.

Conclusions

As before, let

- w = the word length (a positive integer)
- d = dimensionality (a positive integer)
- A = an alphabet (set of symbols)
- D = a dictionary (a set of strings over A of length w)

For fixed w, A and D, with unique word placement, some interesting questions include:

1) For what value of d does the size of the solution space begin to decrease?

2) What is the lowest value of d for which the solution space is known to be empty?

3) What is the highest value of d for which a (w,d) cube is known?

4) To what extent can the empty regions of Tables 1 and 4 be populated through the use of different dictionaries?

5) What are the effects of different hypersolids (non-symmetric, non-cubical, etc.)?

For symmetric hypercubes using the EOWL with w ≥ 6, the answers are known and were presented in previous sections. We show our current answers for w < 6 in the table 5.

Table 5

Known Limits for EOWL Hypercubes

word length w	highest value of d for which (w,d) object found	# words placed for (w,d+1) object after 1 week CPU time
3	17	167/171
4	6	80/84
5	4	**50/70**

For w = 3 with the EOWL, we have found solutions up to d = 17. For d = 18, we were able to place 167 out of 171 words after one week of computation time, so we suspect (3,18) hypercubes exist and perhaps higher dimensional objects as well. For w=4, we have found solutions up to d = 6. For d = 7, we were able to place 80 out of

84 words, so again we suspect objects of higher dimensionality exist, given that the search space is so large.

For w = 5, the question is more problematic. We were only able to place 50 out of 70 words in a week-long search for a (5,5) hypercube. On the other hand, during that time we were only able to search an estimated $1/20^{th}$ of one percent of the solution space. The existence of a (5,5) hypercube for EOWL, or indeed any word list, remains an open question.

References

1. Albert, E. (2012) "The Best 9X9 Square Yet," *Word Ways*: Vol. 24: Issue 4, Article 2

2. Borgmann, D., "Language on Vacation: An Olio of Orthographical Oddities", Charles Scribner & Sons, 1965.

3. The 2012 University of Chicago Scavenger Hunt List: Available online at http://scavhunt.uchicago.edu/scavlist2012.pdf

4. Eckler, A. (2009) "Cubic Word Squares," *Word Ways*: Vol. 42: Issue 1, Article 4

5. EOWL: Available online at http://dreamsteep.com/projects/the-english-open-word-list.html

6. Francis, D. (1971) "From Square to Hyperhypercube," *Word Ways*: Vol. 4: Issue 3, Article 8

7. Garey, M. and Johnson, D., "Computers and Intractability: A Guide to the Theory of NP-Completeness", W.H. Freeman, 1979, ISBN 978-0716710455.

8. Gordon, L. (2012) "Designing a List for Word Squares," *Word Ways*: Vol. 29: Issue 3, Article 4

9. Knuth, D. E. (2012) "5X5X5 Word Cubes By Computer," *Word Ways*: Vol. 26: Issue 2, Article 12.

10. Grant, J. (1978) "Cubism Revisited," *Word Ways*: Vol. 11: Issue 3, Article 1

11. Grant, J. (1979) "More Word Cubes," *Word Ways*: Vol. 12: Issue 2, Article 3

12. Kon, R. (2009) "Solid and Hypersolid Forms," *Word Ways*: Vol. 42: Iss. 4, Article 14

13. Wang Tiling Problem: http://en.wikipedia.org/wiki/Wang_tile

14. Word Square: http://en.wikipedia.org/wiki/Word_square

Mr. Browne and the Dance of Yu:Constructing a Normal Magic Square of Order 3^n

Frank J. Swetz

616 Sandra Ave.

Harrisburg, PA 17109

Abstract

The normal magic square of order three has fascinated viewers and perplexed problem solvers for centuries. Its origins can be traced to ancient China where it was known as the Luoshu [Luo river writing or document] and used as a subject for ritual and numerological manipulation. Over time, the Chinese developed a series of magic squares but these remained ceremonial devices devoid of mathematical theory. In 1917, magic square enthusiast C. A. Browne published a twenty-seven order normal magic square embedded with mystical properties. A mathematical analysis of Mr. Browne's square reveals a relationship to the Luoshu. Through the use of a judicious system of partitioning and repeated iteration of the Luoshu pattern, it appears that for any N, a positive, integer, a normal magic square of order 3^n can be constructed.

An early twentieth century issue of the philosophy journal *The Monist* contained an unusual magic square. In an article entitled " Magic's and Pythagorean Numbers ", a C. A. Browne, no further identification given, produced a magic square supposedly based on the proportions of the Pythagorean Tetractys and supporting number symbolism found in the writings of Plato. It was a normal magic square of order 27, comprised of the integers 1-729, with a central entry of 365 and a magic sum 9855 and it appears in figure 1.

Figure 1

352	381	326	439	468	413	274	303	248	613	642	587	700	729	674	535	564	509	118	147	92	205	234	179	40	69	14
327	353	379	414	440	466	249	275	301	588	614	640	675	701	727	510	536	562	93	119	145	180	206	232	15	41	67
380	325	354	467	412	441	302	247	276	641	586	615	728	673	702	563	508	537	146	91	120	233	178	207	68	13	42
277	306	251	355	384	329	433	462	407	538	567	512	616	645	590	694	723	668	43	72	17	121	150	95	199	228	173
252	278	304	330	356	382	408	434	460	513	539	565	591	617	643	669	695	721	18	44	70	96	122	148	174	200	226
305	250	279	383	328	357	461	406	435	566	511	540	644	589	618	722	667	696	71	16	45	149	94	123	227	172	201
436	465	410	271	300	245	358	387	332	697	726	671	532	561	606	619	648	593	202	231	176	37	66	11	124	153	98
411	437	463	246	272	298	333	359	385	672	698	724	507	533	559	394	620	646	177	203	229	12	38	64	99	125	151
464	409	438	299	244	273	386	331	360	725	670	699	560	505	534	647	592	621	230	175	204	65	10	39	152	97	126
127	156	101	214	243	188	49	78	23	361	390	335	448	477	422	283	312	257	593	624	569	682	711	656	317	346	491
102	128	154	189	215	241	24	50	76	336	362	388	423	449	475	258	284	310	570	596	622	657	683	709	492	518	344
155	100	129	242	187	216	77	22	51	389	334	363	476	421	450	311	256	285	623	568	597	710	655	684	545	490	519
52	81	26	130	159	106	208	237	182	286	315	260	364	393	338	442	471	416	520	549	494	598	627	572	676	705	650
27	53	79	105	131	157	183	209	235	261	287	313	339	365	391	417	443	469	495	521	547	573	599	625	651	677	703
80	25	54	158	103	132	236	181	210	314	259	288	392	337	366	470	415	444	548	493	522	626	571	600	704	649	678
211	240	185	46	75	20	133	162	107	443	474	419	280	309	254	387	396	341	679	708	653	514	543	488	601	630	575
186	212	238	21	47	73	108	134	160	420	446	472	255	281	307	342	368	394	654	680	706	489	515	541	376	602	628
239	184	213	74	19	48	161	106	135	473	418	447	308	253	282	395	340	369	707	652	681	542	487	516	629	576	603
604	633	578	691	720	665	526	555	500	109	138	83	196	225	170	31	60	5	370	399	344	457	486	431	292	321	266
579	605	631	666	692	718	501	527	553	84	110	136	171	197	223	6	32	58	345	371	397	432	458	484	267	293	319
632	577	606	719	664	693	554	499	528	137	82	111	224	169	198	59	4	33	398	343	372	485	430	459	320	265	294
529	558	503	607	636	581	685	714	659	34	63	8	112	141	86	190	219	164	295	324	269	373	402	347	451	480	425
504	530	556	582	608	634	660	686	712	9	35	61	87	113	139	165	191	217	270	296	322	348	374	400	426	452	478
557	502	531	635	580	609	713	658	687	62	7	36	140	85	114	218	163	192	323	268	297	401	346	375	479	424	453
688	717	662	523	552	497	610	639	584	193	222	167	28	57	2	115	144	89	454	483	428	289	318	263	376	405	350
663	689	715	498	524	550	585	611	637	168	194	220	3	29	55	90	116	142	429	455	481	264	290	316	351	377	403
716	661	690	551	496	525	638	583	612	221	166	195	56	1	30	143	88	117	482	427	456	317	262	291	404	349	378

While the numerological associations of the square are interesting, the real mystery lies in its construction. Paul Carus, mathematician, philosopher, and then founding editor of *The Monist* and himself a magic square enthusiast, commented on the uniqueness of the square [1]. In analyzing it , Carus noted that it was composite and that its construction depended on an extension of the 3 x 3 natural magic square [Luoshu] devised by a repeated use of the Knight's movement staircase technique. My more recent investigation of this square reveals an interative pattern that can, I believe, be used to construct any magic square of order 3^n for all n.

The Luoshu

The three by three configuration of natural numbers known as the Luoshu is the oldest recorded magic square. Its origins can be traced back to fourth century BCE China. See Figure 2.

Figure 2

4	9	2
3	5	7
8	1	6

The square's use as a cosmological and ritual device within Chinese ritual traditions was firmly established from that early time onwards [2]. Later Daoist authors formalized a technique for constructing this magic square. They presented the method as a ritual movement, a dance, whose graphic depiction served as a visual algorithm [3]. This algorithm is a network of line segments whose vertices and ordering indicate the positions of the natural numbers within the Luoshu configuration. The resulting ritual movement was called the *Yubu* or "the dance of Yu" (Yu being a legendary emperor). See figure 3. The Chinese derivation of the Luoshu proceeds from the natural square order three where the numbers are arranged employing a Chinese lexographical ordering i.e. right to left, top to bottom. See figure 4. Then following the directions of the Yubu, the elements of the square can be considered acted upon by a permutation, μ, which maps them into the Luoshu configuration where:

$$\mu = \begin{pmatrix} 1 & 2 & 3 & 4 & 5 & 6 & 7 & 8 & 9 \\ 6 & 1 & 8 & 7 & 5 & 3 & 2 & 9 & 4 \end{pmatrix}$$

Furthermore, let the geometric arrangement of cells as designated in Figure 4 define a partition , P_k , that acts upon any square array of numbers of order 3^k such that it physically divides the array into nine equal cells each containing sub arrays of order 3^{k-1} sequenced as shown in Figure 4, that is, the first sub array occupies the upper right cell and the next one will be directly below it and so on. Then given any natural number square of order 3^n , $N(3^n)$, n alternate sets of partitions and permutations will transform the square into a magic square of order 3^n , $M(3^n)$.

Figure 3

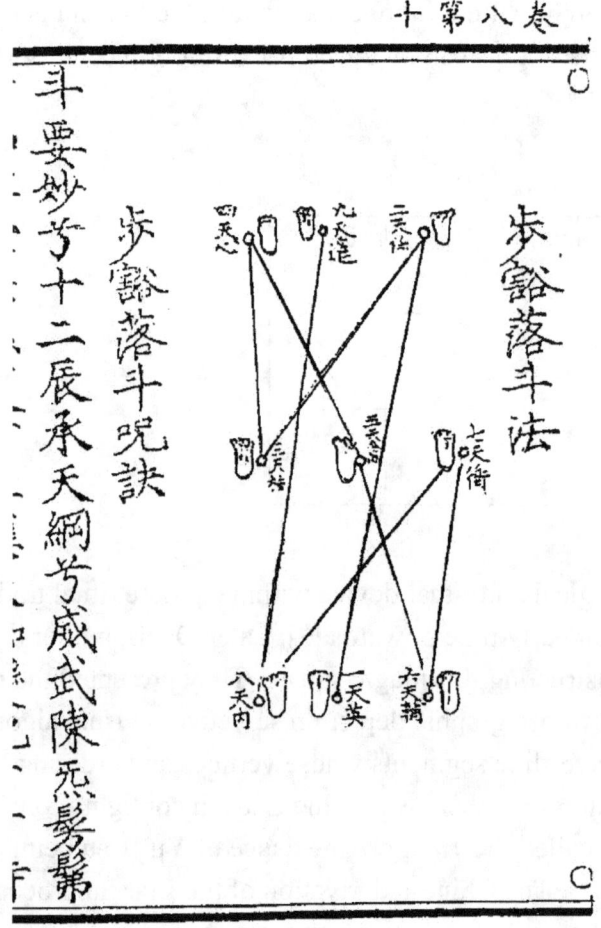

Figure 4

7	4	1
8	5	2
9	6	3

Thus,

$$M(3^k) = (\mu P_k [N(3^k)]) \big|_k$$

here the index k = n, n-1,...., 1, since the partition P_k operates on the largest number square first.

An Example, M (3^2)

For simplicity, consider the use of this algorithm in constructing the magic square of order nine. First construct the natural number square of order 9 using the Chinese lexographical ordering, then apply P_2 to the square. The resulting division for the first column is shown by the schema in figure 5.

Figure 5

73	46 37 28	19 10 1 20 11 2 21 12 3
		22 13 4 23 14 5 24 15 6
81		25 16 7 26 17 8 27 18 9

Then apply μ (P_2) to this square. For simplicity just consider the transformation of the numbers in the cells of the rightmost column only, shown in figure 6.

Now P_1 and μ (P_1) are applied to each 3x3 array, converting each into a magic square, as can be seen in figure 7.

Figure 6

		22 13 4
		23 14 5
		24 15 6
25 16 7		
26 17 8		
27 18 9		
	19 10 1	
	20 11 2	
	21 12 3	

Figure 7

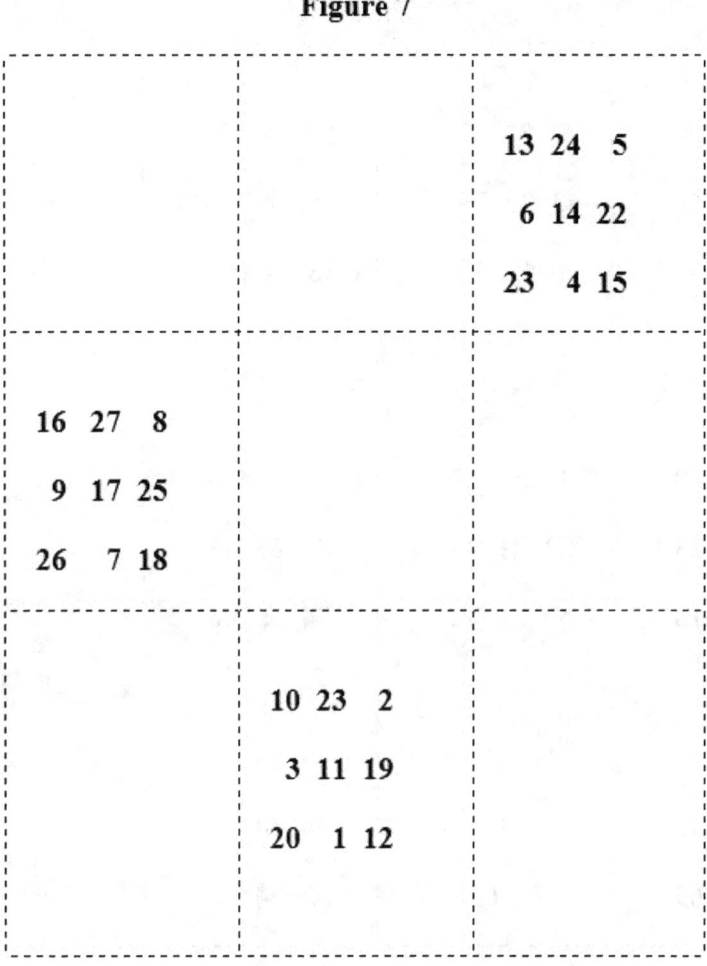

When the algorithm is applied to all the elements of N (3^2), a ninth order magic square results, as seen in figure 8.

Figure 8

37	48	29	70	81	62	13	24	5
30	38	46	63	71	79	6	14	22
47	28	39	80	61	72	23	4	15
16	27	8	40	51	32	64	75	56
9	17	25	33	41	49	57	65	73
26	7	18	50	31	42	74	55	66
67	78	59	10	21	2	43	54	35
60	68	76	3	11	9	36	44	52
77	58	69	20	1	12	53	34	45

Marie Baldys, a mathematical games enthusiast, of Harrisburg, PA, employing a JAVA program applied the algorithm to N (3^3) and obtained Mr. Browne's magic square. She also proceeded further successfully obtaining the magic squares for N (3^4) and N (3^5). Some of these results can be found on her website [4]. I urge interested readers to extent the results further.

Final Comments

The early Chinese did not overtly pursue a mathematical study of magic squares. The squares were mere ritual entities with spiritual rather than intellectual appeal. The earliest existing compilation of Chinese magic squares appears in the Sung Dynasty with the publication of Yang Hui's *Yang Hui suanfa*,[Yang Hui's Methods of Computation], (1275) [5].Yang considers squares up to the 10th order but views them as arithmetical curiosities. About thirty years later scholar Ding Yidong published *Dayan suoyin,* a treatise on the ritual use of magic squares[6]. Both Yang and Ding gave examples of ninth order magic squares that were influenced by Yubu but neither adhered to the algorithm above. Modern surveys of magic squares and their construction show no evidence of Yubu techniques [7] .The particular scheme Mr. Browne employed to arrive at his *Monist* square may never be known but it's existence has inspired examination and, so it seems, revealed a link to the ancient "dance of Yu".

References

1. W.Andrews. *Magic Squares and Cubes*, Dover publishing Co. New York, pp.146-162,1960 (Reprint of 1917 edition.).

2. F. Swetz.: *Legacy of the Luoshu,The 4000 Year Search for the Meaning Of the Magic Square of Order Three*, Wellesley, MA: A.K.Peters, 2008.

3. P. Anderson. The Practice of Bugang, *Cahiers d'Extreme-Asie*, pp. 15-53, (1989-90).

4. http://www.tomargames.com/tmLuoshu/luoshu.html .

5. L.L.Yong. *A Critical Study of Yang Hui Suan Fa:A Thirteenth-Century Mathematical Treatise*, Singapore:Singapore University Press,1977.

6. P.Y.Ho, Chinese Science:the Traditional Chinese View, *Bulletin of the School of Oriental and African Studies University of London*, 54:506-519, (1991).

7. C.A.Pickover.*The Zen of Magic Suares,Circles ,and Stars*, Princeton: Princeton University Press, 2002.

Geometry and Design of Equiangular Spirals

Kostantinos Myrianthis

58, Zan Moreas str., Athens, P.C.15231,Greece

myrian@ath.forthnet.gr

Abstract

In an equiangular spiral, "the whorls continually increase in breadth and do so in a steady and unchanging ratio... It follows that the sectors cut out by successive radii, at equal vectorial angles, are similar to one another in every respect and that the figure may be conceived as growing continuously without ever changing its shape the while" as stated by Sir D'Arcy W. Thompson and quoted in [1, p.125]. I was fascinated since my early years with the shape of spirals and all their versions in nature. The mathematical modeling of them became a very attractive topic of study and research for me and more specifically, the geometrical conditions under which any quadrangle or triangle can be fitted into similar copies of itself and form an equiangular spiral. This formation gives the impression of a digital form of spiral, where every digit is a triangle or quadrangle following similarity laws, which can allow a multiplicity of design capabilities. The essence of this work appears in the present article and is related with the geometry and the design characteristics of equiangular spirals.

Introduction

In figure 1a we have the branch $A_{0,0} A_{0,1} A_{0,2} A_{0,3}$ ... defined by the equally spaced (by angle φ) rays $SA_{0,0}$, $SA_{0,1}$, $SA_{0,2}$... from the centre S of the spiral. The triangles $\Delta A_{0,0} A_{0,1} S$, $\Delta A_{0,1} A_{0,2} S$, $\Delta A_{0,2} A_{0,3} S$,... are similar with a ratio of similarity κ. As shown in Appendix 1, part1, $(A_{0,0} A_{0,1} \wedge A_{0,1} A_{0,2}) = \varphi$. The angle σ can be expressed in relation to φ and κ (as shown in Appendix 1, part 2):

Figure 1a

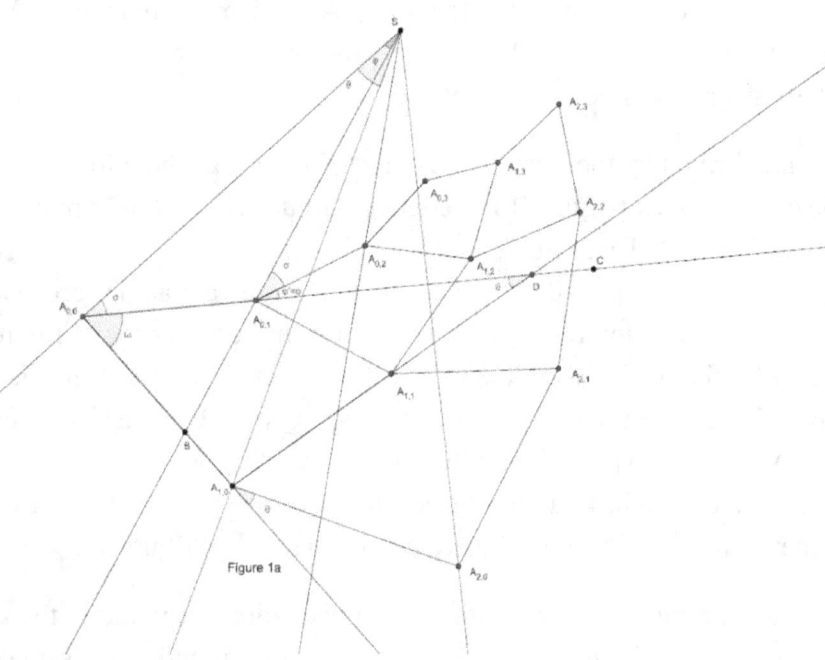

Figure 1a

$$(1) \qquad \tan \sigma = \kappa \sin \varphi \ / \ (1 - \kappa \cos \varphi)$$

Rule 1: It is obvious that given the position and size of 2 consecutive segments of a spiral branch so that φ, σ and κ can be calculated, then the position of S can be determined.

All the branches $A_{0,0} A_{0,1} A_{0,2} A_{0,3}$..., $A_{1,0} A_{1,1} A_{1,2} A_{1,3}$..., $A_{2,0} A_{2,1} A_{2,2} A_{2,3}$, can be named $B\kappa_0 = A_{0,0} A_{0,1} A_{0,2} A_{0,3}$..., $B\kappa_1 = A_{1,0} A_{1,1} A_{1,2} A_{1,3}$..., $B\kappa_2 = A_{2,0} A_{2,1} A_{2,2} A_{2,3}$..., or generally $B\kappa_i$ where i = 0, 1, 2, ... in the direction of rotating of similarity ratio κ for simplicity.

In the same figure 1a we have the branch $A_{0,0} A_{1,0} A_{1,1} A_{1,2}$..., defined by the equally spaced (by angle θ) rays $SA_{0,0}$, $SA_{1,0}$, $SA_{2,0}$... from the centre S of the spiral. The triangles $\Delta A_{0,0} A_{1,0} S$, $\Delta A_{1,0} A_{2,0} S$, $\Delta A_{2,0} A_{3,0} S$,... are similar with a ratio of similarity λ. As with the case of similarity ratio κ, $(A_{0,0} A_{1,0} \wedge A_{1,0} A_{2,0}) = \theta$ and:

(2) $\tan(\omega+\sigma) = \lambda\sin\theta/(1 - \lambda\cos\theta)$

As with the case of similarity ratio κ all the branches $A_{0,0}A_{1,0}A_{2,0}A_{3,0}...$, $A_{0,1}A_{1,1}A_{2,1}A_{3,1}..., A_{0,2}A_{1,2}A_{2,2}A_{3,2}$, can be named $B\lambda_i$ where $i = 0, 1, 2, \ldots$ in the direction of rotating of similarity ratio λ for simplicity.

Rule 2: If the angle σ is less than $\pi/2-\varphi/2$, ($SA_{0,0} > SA_{0,1}$), then the branches $B\kappa_i$ rotate anticlockwise and converge to S, otherwise ($SA_{0,0} < SA_{0,1}$) they rotate clockwise and converge to S (for $\sigma =\pi/2-\varphi/2$ the triangle $\Delta A_{0,0}A_{0,1}S$ is isosceles and these branches do not converge). Similarly, if the angle ($\omega + \sigma$) is less than $\pi/2-\theta/2$, ($SA_{0,0} < SA_{1,0}$), then the branches $B\lambda_i$ rotate anticlockwise and converge to S, otherwise ($SA_{0,0} > SA_{1,0}$) they rotate clockwise and converge to S (for ($\omega+\sigma$) = $\pi/2-\theta/2$, these branches do not converge).

The quadrangles which are formed by the branches $B\kappa_i$ and $B\lambda i$ (such as the initial one $A_{0,0}A_{0,1}A_{1,1}A_{1,0}$) can be either convex (figure 1a) or concave (figure 1b) and their relevant angles are calculated in Appendix 1, part 4. These quadrangles can become triangles in two cases, first when $\omega=\varphi$ (figure 1c) and second when $\omega=0$ (figure 1d) and their relevant angles are also calculated in Appendix 1, part 4. It is interesting to note that for the case of $\omega=0$ of figure 1d, the triangles $\Delta A_{0,0}A_{1,0}S$, $\Delta A_{1,0}A_{2,0}S,...$ which form the ratio λ do not have a visual clear role. As shown in Appendix 1 part 5, we have the ratio $\lambda'= A_{0,2}A_{1,1}/A_{1,1}A_{2,0} = \kappa/\lambda$ related to the case of figure 1d, the triangles $\Delta A_{0,2}A_{1,1}S$, $\Delta A_{1,1}A_{2,0}S$ and the angle $\theta' = {}^{\wedge}A_{0,2}SA_{1,1}= {}^{\wedge}A_{1,1}SA_{2,0}= \theta - \varphi$, where $\theta > \varphi$ and these parameters do have a visual role (the case where $\theta < \varphi$ can be treated accordingly). The parameters θ' and λ' can be treated just like θ and λ at figure 1c.

Starting from any quadrangle, for example the initial one, any quadrangle found in the κ direction has a ratio of similarity κ with the two adjacent quadrangles found in the same direction, therefore the quadrangles $A_{0,0}A_{0,1}A_{1,1}A_{1,0}$ and $A_{0,1}A_{0,2}A_{1,2}A_{1,1}$ have a ratio of similarity κ. Similarly, any quadrangle found in the λ direction has a ratio of similarity λ with the two adjacent quadrangles found in the same direction, therefore the quadrangles $A_{0,0}A_{0,1}A_{1,1}A_{1,0}$ and $A_{1,0}A_{1,1}A_{2,1}A_{2,0}$ have a ratio of similarity λ.

Also in figure 1a the branches $B\kappa_i$ are anticlockwise and $B\lambda_i$ are clockwise, so the branches in the κ direction are contra-rotating with respect to the branches of the λ direction. In the figures 1b, 1c, 1d, the branches $B\kappa_i$ and $B\lambda_i$ are co-rotating.

The basic parameters which define the geometry of a spiral system (with quadrangles or triangles) are ω, φ, θ, κ and λ. Given any four of them, the fifth can be calculated. Given these parameters and a segment of a spiral branch of the system which will be assumed as the starting segment such as $A_{0,0}A_{0,1}$ or $A_{0,0}A_{1,0}$, the spiral system can be constructed having a common starting ray $SA_{0,0}$ for both the first branch $B\kappa_0$ of the κ direction and the first branch $B\lambda_0$ of the λ direction.

Figure 1b

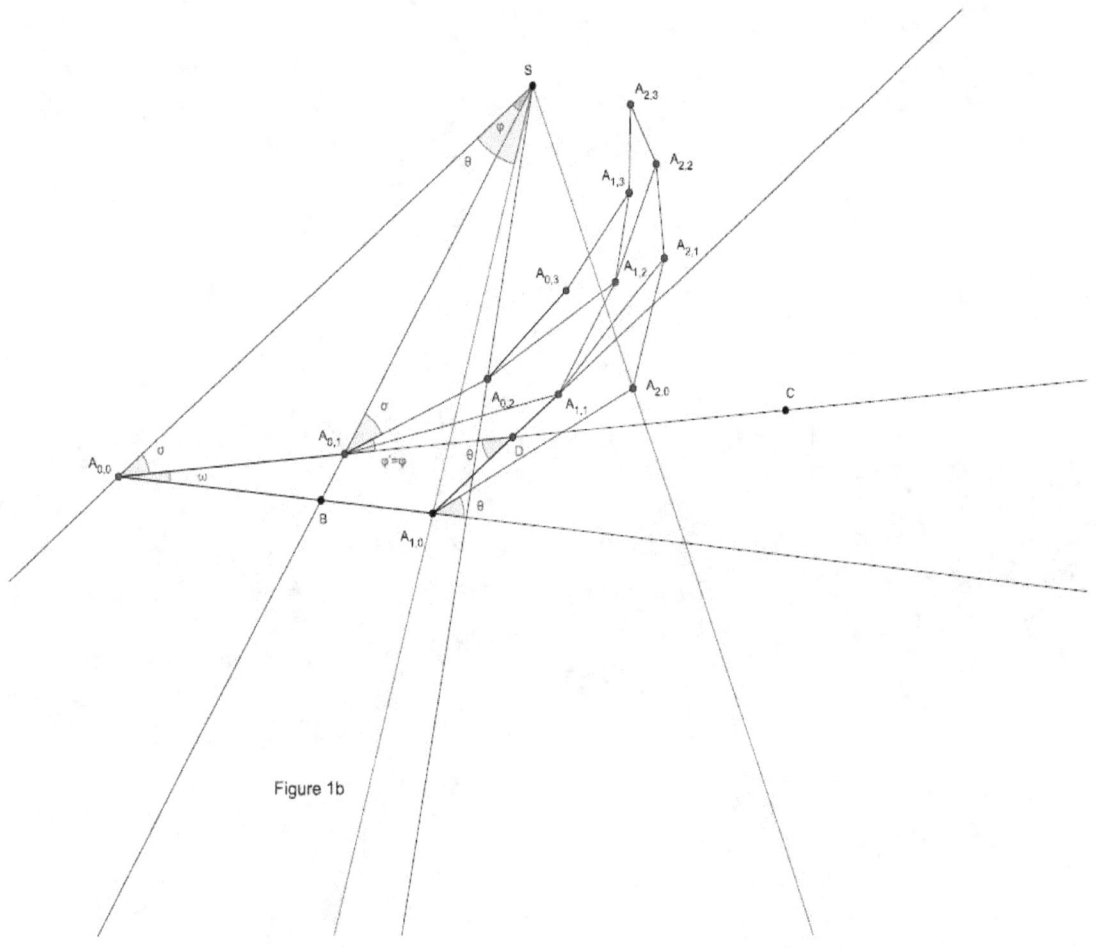

Figure 1b

Any arbitrary quadrangle (or triangle) which can be fitted into similar copies of itself, can produce a spiral system with branches of 2 types (one related with angle φ and ratio κ and another one related with angle θ and ratio λ, as in figure 1a. The result usually overlaps itself or has cracks that cannot be filled, taking into account the related 2-D plane of the spiral system. These types of spiral systems (which can be called open), have an infinite number of branches in both κ and λ directions.

Figure 1c

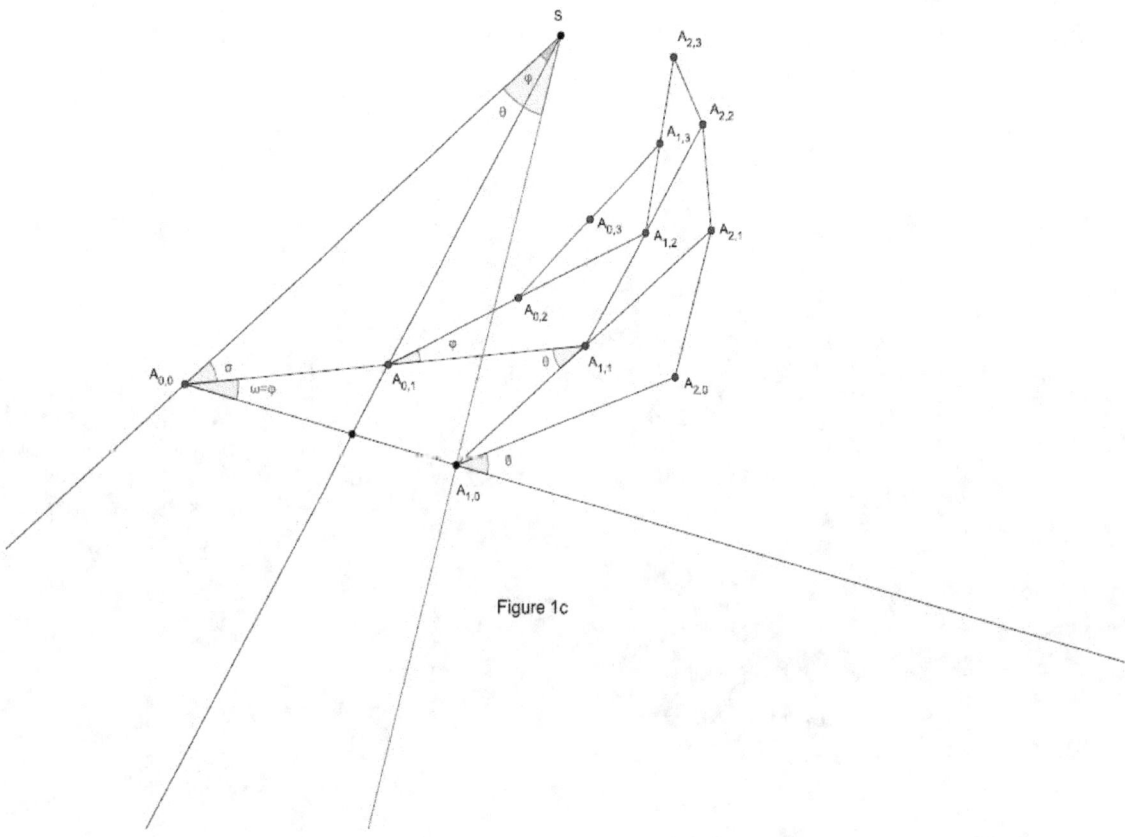

Figure 1c

The intention of the present work is to provide the necessary conditions and parameter values using Euclidean geometry and algebra in order to produce a planar closed spiral system without cracks and overlaps, a closed spiral tiling system if one considers that quadrangles (or triangles) as tiles. In section 2, closed quadrangular systems are examined, whereas in sections 3 and 4 the two types of closed triangular systems ($\omega = \varphi$ and $\omega = 0$) are examined. All the possible cases of the equivalence of triangular spirals are analyzed in section 5. Finally, in section 6, the concept of divergence angle ([4, p.20]) is analyzed.

It is interesting to note that the topology of spiral systems is closely related to phyllotaxis theory as shown in [3] and [4] where complex exponential functions produce quadrangular and triangular closed spiral systems, which find wonderful applications in nature. A pure Euclidean approach such as the one of the present work might give a simpler and more practical point of view with enough interest for future development.

72

Figure 1d

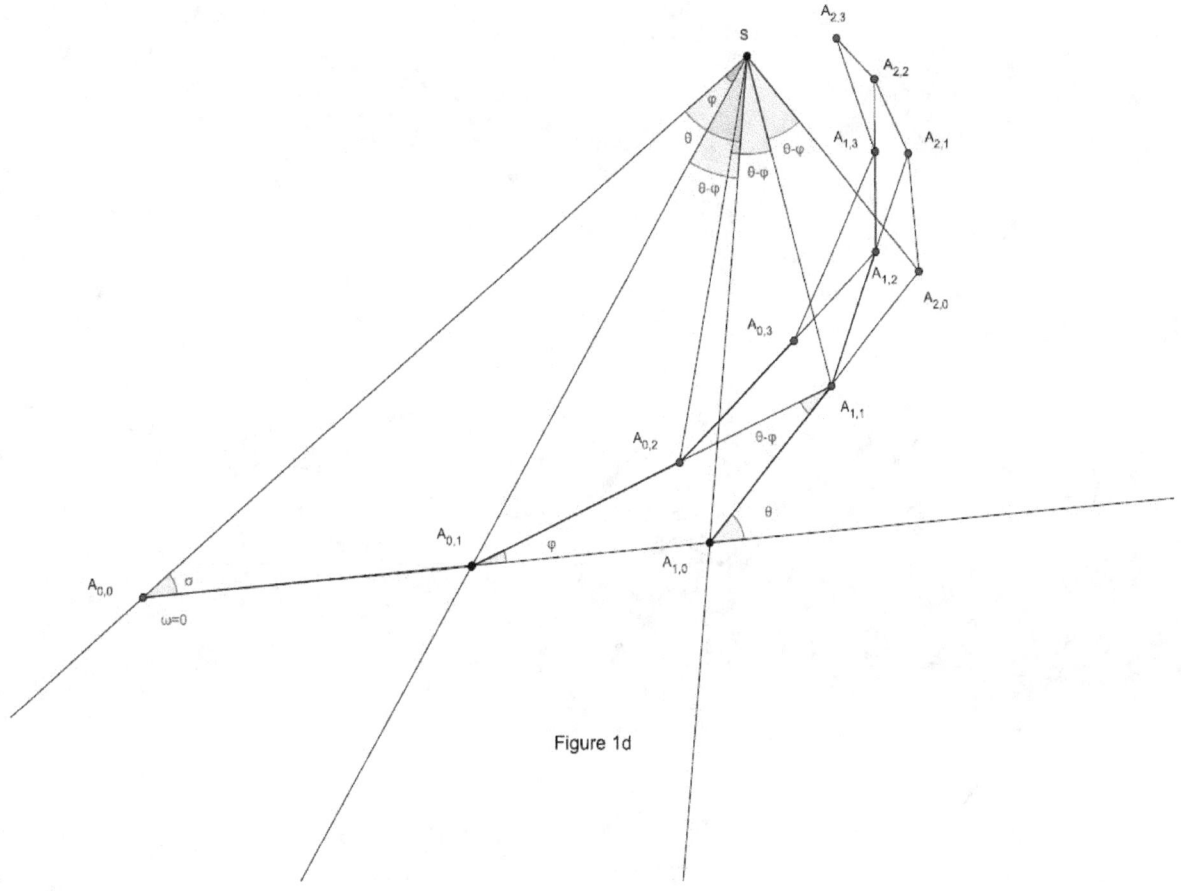

Figure 1d

Closed Quadrangular Spiral Systems

Let $A_{0,0}A_{0,1}A_{1,1}A_{1,0}$ be one of the quadrangles shown in figure 2a, which according to the numbering system of the vertices of the quadrangles (or spiral branches) is the first one. Two branches appear in this figure, $B\lambda_0$ and $B\lambda_1$ in the λ direction with 14 vertices each (0,..,13), and two branches $B\kappa_0$ and $B\kappa_1$ in the κ direction with 4 vertices each (0,...,3), which co-rotate, (plus 12 branches $B\kappa_i$ where i = 2, … 13 in the κ direction with 2 vertices each 0,..,1). As it is shown in Appendix 1, part 3, $(A_{0,0}A_{0,1} \wedge A_{1,0}A_{1,1}) = \theta$, therefore $(A_{0,0}A_{0,1} \wedge A_{13,0}A_{13,1}) = 13\theta$. Also it is obvious that $(A_{0,0}A_{0,1} \wedge A_{0,2}A_{0,3}) = 2\varphi$.

Obviously the open spiral system presented in figure 2a has a gap or it has overlaps if all the quadrangles are extended in the κ and λ directions. The initial branches $B\kappa_0$ and $B\lambda_0$ start from the same vertex $A_{0,0}$ and they never meet again at any vertex of the spiral, also there is an infinite number of branches in both κ and λ directions.

73

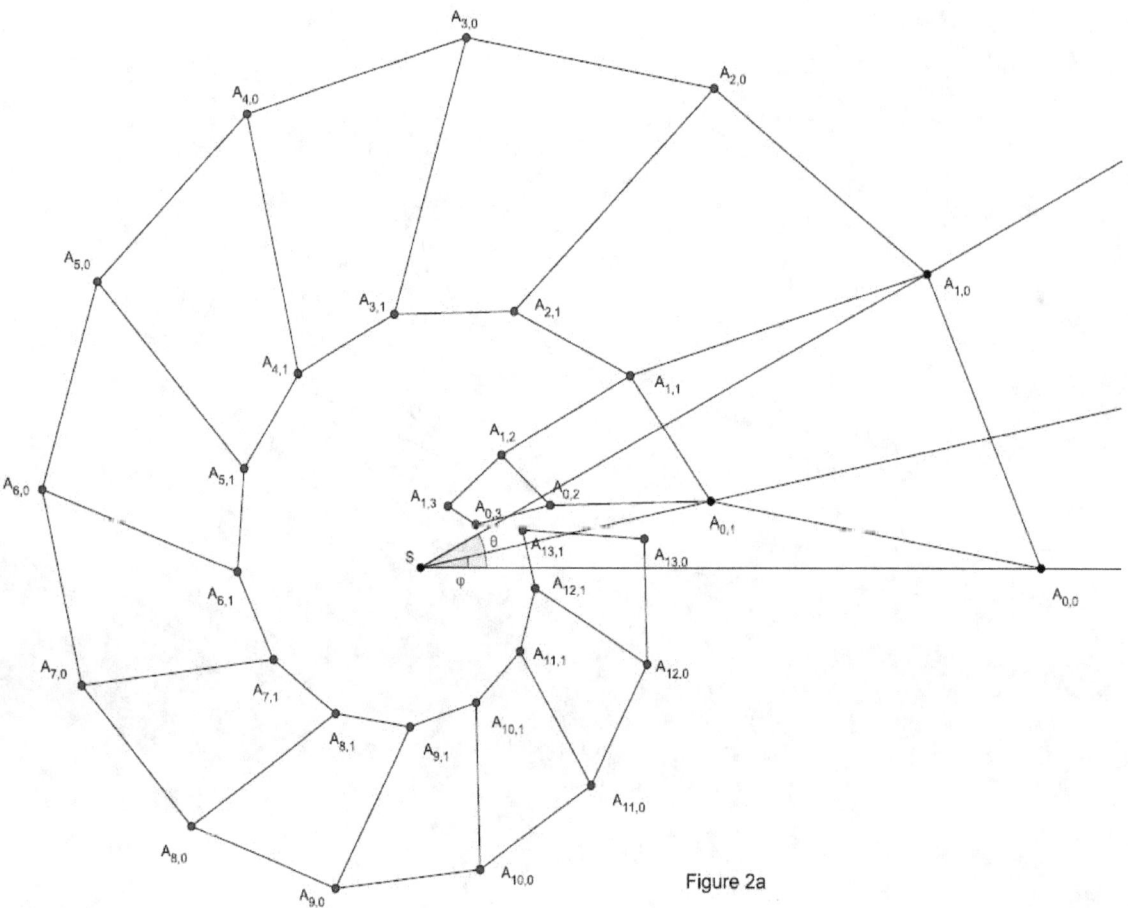

Figure 2a

In figure 2b we have a spiral system of which the branches of κ and λ directions co-rotate, where there is no gap or overlap, so it can be considered as a closed spiral system. Because of this property, the numbering process of the vertices (or nodes) of the quadrangles of the spiral system has to be different from the one of figure 2a. If we consider the segment $A_{13,1}A_{13,0}$ of figure 2a and assume that the spiral system of figure 2a is somehow transformed into the spiral system of figure 2b, it is as if this segment has moved and coincided with the segment $A_{0,2}A_{0,1}$ of figure 2a in order to produce the segment $A_{26,2}A_{13,1}$ of figure 2b. In other words, in figure 2b the initial branch $B\lambda_0$ starts from the vertex $A_{0,0}$ and meets for the first time the initial branch $B\kappa_0$ (which also starts from the vertex $A_{0,0}$) at the vertex $A_{13,1}$, which is the first vertex or point after the common start of the two branches. Given that a typical vertex of the spiral is $A_{i,j}$, in this case we have i = 13 and j = 1. Similarly, the two branches meet for the second time at the vertex $A_{26,2}$, and so on. In figure 2b we have one branch $B\lambda_0$ in the λ direction and 13 branches $B\kappa_i$ where i = 0, … ,12 in the κ direction.

Figure 2b

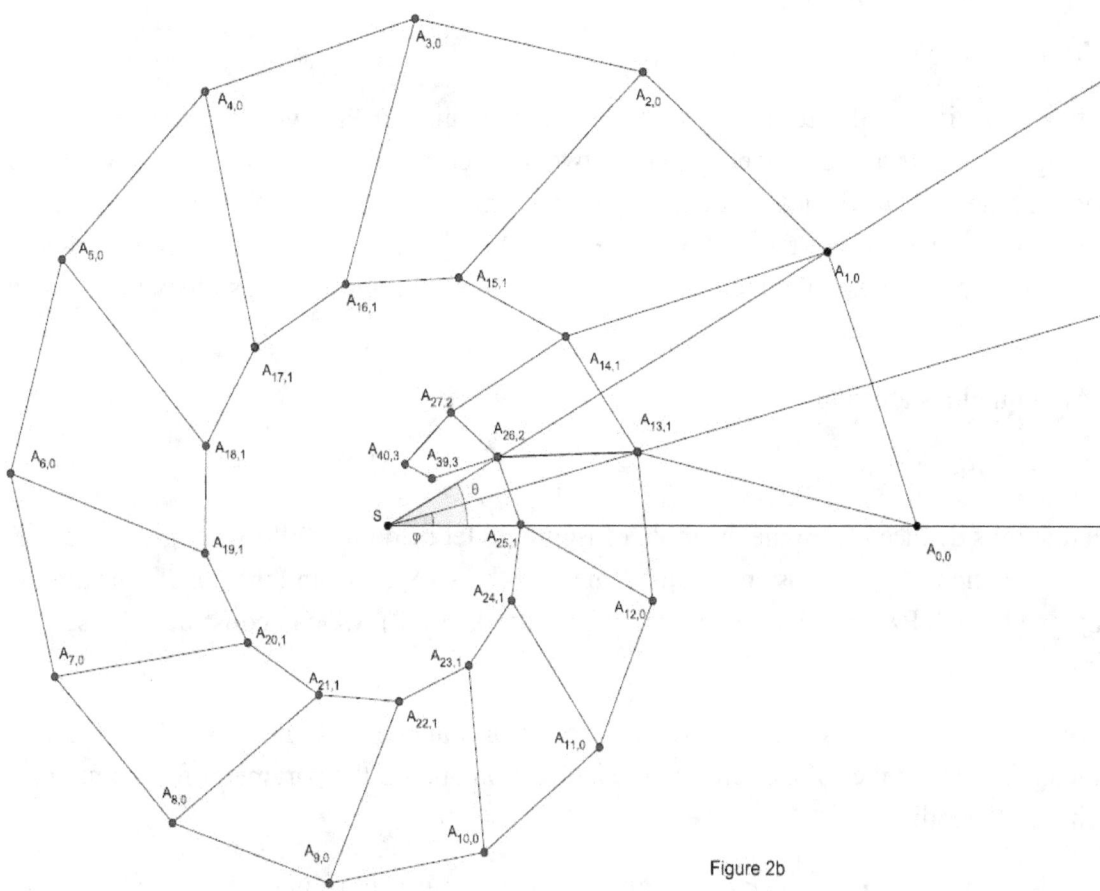

Figure 2b

We have $(A_{0,0}A_{13,1} \wedge A_{13,1}A_{26,2}) = \varphi$ and $(A_{0,0}A_{13,1} \wedge A_{1,0}A_{14,1}) = \theta$ (Appendix 1 part 1 and part 3). From the above and from figure 2b, we get:

(3) $13\theta = 2\pi + \varphi$

This is better understood by considering a "traveling" segment which starts from the initial segment's $A_{13,1}A_{0,0}$ position, goes to $A_{14,1}A_{1,0}$ (first step) in the λ direction, then to $A_{15,1}A_{2,0}$ (second step) and so on, covering in 13 steps all positions in order to "arrive" finally to $A_{26,2}A_{13,1}$ (which forms an angle φ with the segment $A_{13,1}A_{0,0}$). This "traveling" segment rotates at a total angle 13θ in order to arrive at the final position. In the same way another "traveling" segment starts from the same initial segment $A_{13,1}A_{0,0}$ in the κ direction and goes to $A_{26,2}A_{13,1}$ in one step, rotating at an angle of φ. These two "traveling" segments in order to meet together and appear as one segment, apart from rotating in order to have the same final orientation, they have to have the same size. Because of the branch $B\kappa_0$ and the ratio of similarity ratio κ, we have:

$A_{26,2}A_{13,1} = \kappa A_{13,1}$

75

Also, due to the fact that any quadrangle found in the λ direction has a ratio of similarity λ with the two adjacent quadrangles found in the same direction and because the first "traveling" segment "travels" in 13 steps in that direction, we have:

$$A_{26,2}A_{13,1} = \lambda^{13}A_{13,1}$$

Apart from the issues of size and rotation, there is also another fact which makes the two "traveling" segments to meet together. These two segments at their final position have the same orientation, the same size and additionally have to form with the center S of the spiral a triangle with a known size and orientation of one side ($A_{26,2}A_{13,1}$, their size when they meet) and angles φ, σ, $\pi - \varphi - \sigma$ so there is only one triangle satisfying these requirements, so they meet together and appear as one segment.

The two equations above give us

(4) $\qquad \lambda - {}^{13}\sqrt{\kappa}$

The equations deduced from the example of figure 2b let us do the following: given the defining parameters which are φ, κ plus an initial segment such as $A_{0,0}A_{13,1}$ and the conditions that there is only one branch $B\lambda_0$ and thirteen branches $B\kappa_i$ ($i = 0, \ldots, 12$), we can construct a closed quadrangle spiral system.

Rule 3: In a closed spiral system where the number of branches is finite, we define as n, the parameter for the number of branches in the λ direction and m, the parameter for the number of branches in the κ direction.

In the case of figure 2b, we have $n = 13$ and $m = 1$. In figure 2c we have the same defining parameters as in figure 2b, except for $m = 2$, which is the number of branches in the κ direction. Thus, the initial branch $B\lambda_0$ meets the initial branch $B\kappa_0$ at vertex $A_{13,2}$ ($A_{i,j}$ where $i = 0 + 13$, $j = 0 + 2$) after their common start and in a similar way the second branch $B\lambda_1$ and the initial branch $B\kappa_0$ meet at vertex $A_{13,3}$ ($A_{i,j}$ where $i = 0 + 13$, $j = 1 + 2$) after their common start at vertex $A_{0,1}$. The branches $B\lambda_0$ and $B\kappa_0$ meet again at point $A_{26,4}$ ($A_{i,j}$ where $i = 0 + 2 \times 13$, $j = 0 + 2 \times 2$), the branches $B\lambda_1$ and $B\kappa_0$ at point $A_{26,5}$ ($A_{i,j}$ where $i = 0 + 2 \times 13$, $j = 1 + 2 \times 2$) and so on. This is how the vertices numbering system of the closed spiral systems works.

The equations 3 and 4 become respectively: $13\theta = 2\pi + 2\varphi$ and $\lambda = {}^{13}\sqrt{\kappa^2}$. Their general form is:

(5) $\qquad\qquad\qquad n\theta - m\varphi = 2\pi \Leftrightarrow n\theta = 2\pi + m\varphi$

(6) $\qquad\qquad\qquad \lambda^n = \kappa^m \Leftrightarrow \lambda = {}^n\sqrt{\kappa^m}$

It has to be noted that the values of κ and λ which are related to equation 6 have to be either greater or less than one (usually $\kappa < 1$, so we have to have $\lambda < 1$ as well), so the equations 1, 2, 8,

17 have to be treated accordingly (if the branches $B\kappa_i$ and $B\lambda_i$ are contra-rotating, the reciprocal expressions of either κ or λ have to be used).

Figure 2c

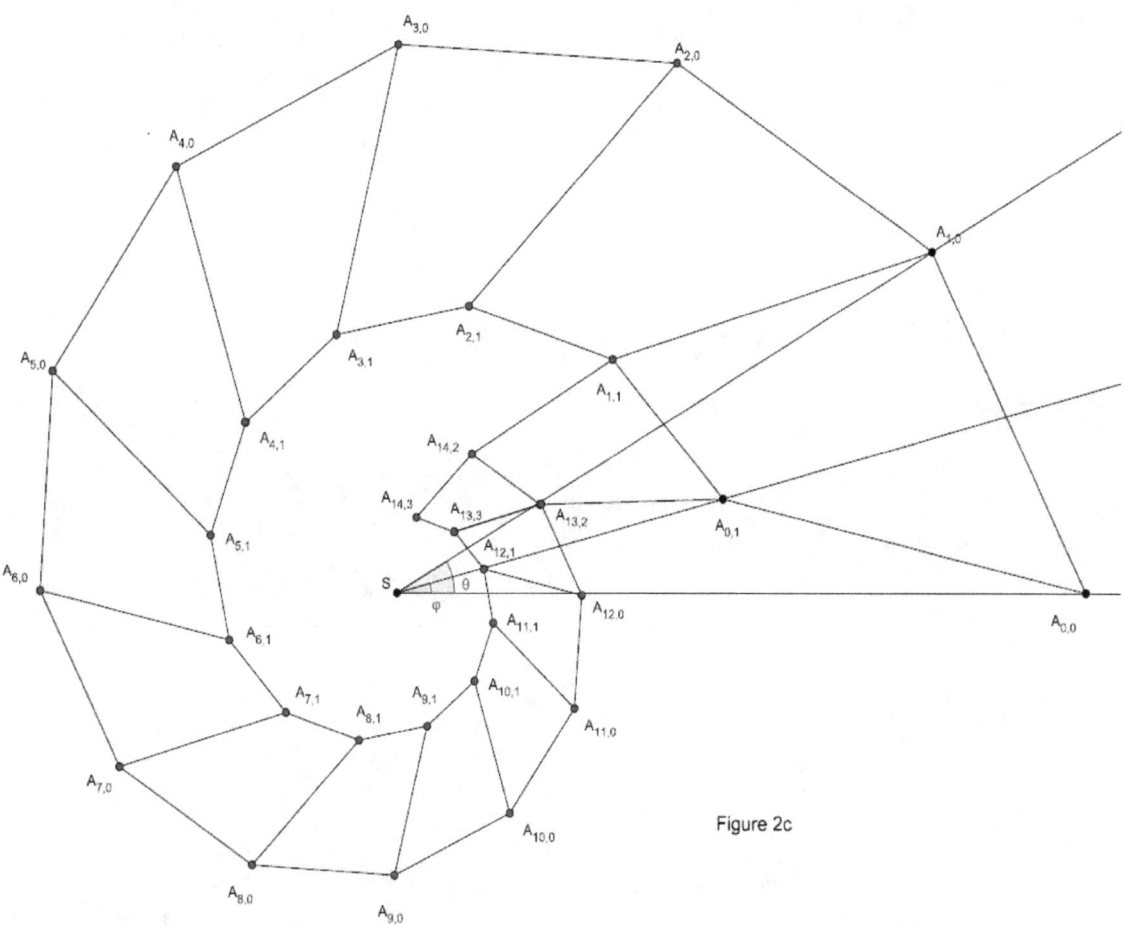

Figure 2c

 In figure 2d we have a spiral system of which the branches of κ and λ directions contra-rotate, where n = 11 and m = 1. The only difference is in the equation 5 (general form) which becomes:

(7) $n\theta + m\varphi = 2\pi \Leftrightarrow n\theta = 2\pi - m\varphi$

 The above equation states that the "traveling" segment in figure 2d (which starts from the initial segment's $A_{13,1}A_{0,0}$ position, as in figure 2b) rotates covering a total angle 11θ (n = 11) which is equal to 2π minus the angle φ, (since φ is the angle between the segments $A_{0,0}A_{11,1}$ and $A_{11,1}A_{22,2}$ and m = 1). The equation 7 is necessary but not sufficient in order to have a closed quadrangular spiral system with contra-rotating branches, so the right combination of conditions of rule 2 has to apply, which in this case is:

$\sigma < \pi/2 - \varphi/2$ and $(\omega + \sigma) > \pi/2 - \theta/2$ or

77

$\sigma > \pi/2 - \varphi/2$ and $(\omega + \sigma) < \pi/2 - \theta/2$

Figure 2d

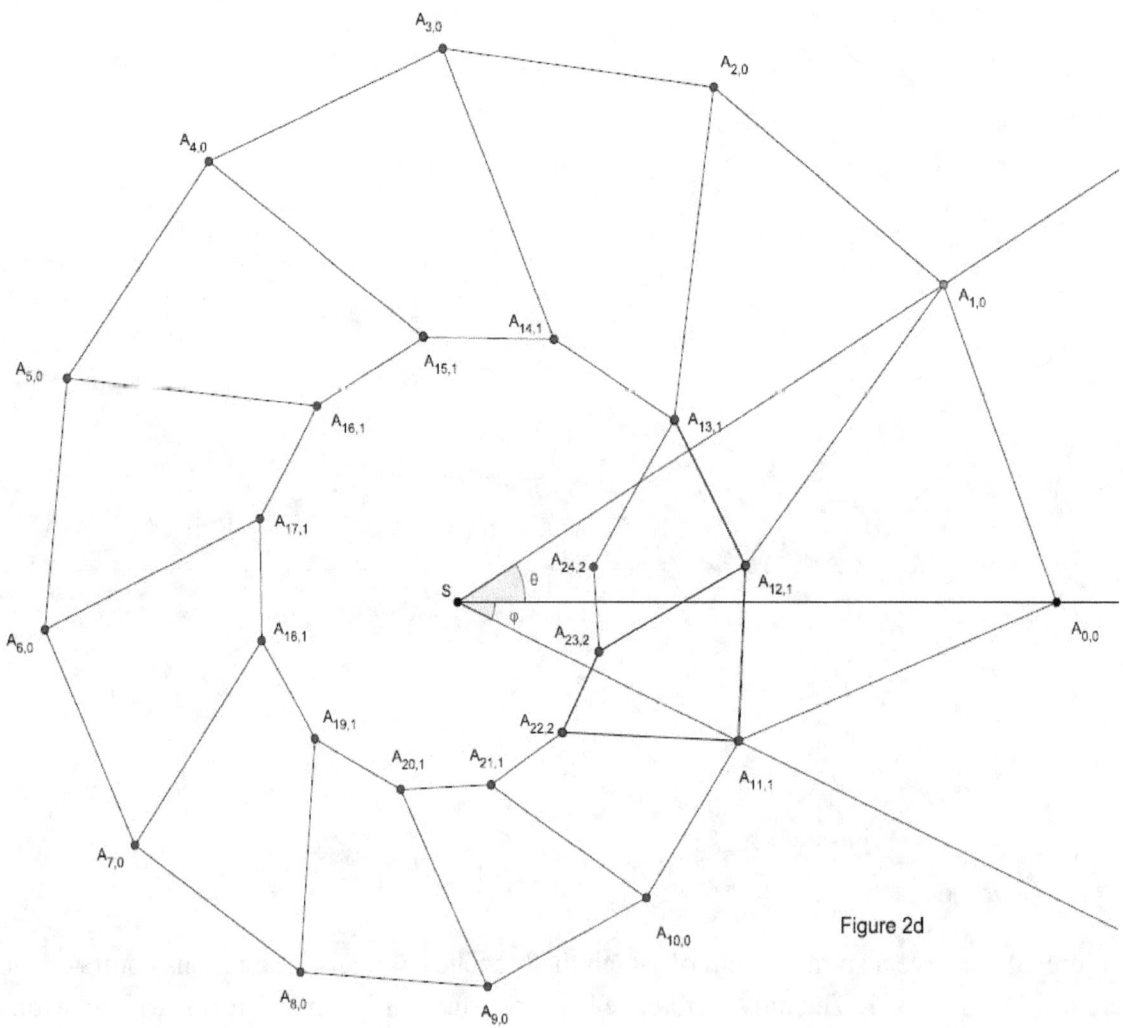

Figure 2d

The following parameters belong to an example where equation 7 holds but is not sufficient, $\varphi = 30\,\text{degr}$, $n = 6$, $m = 4$ and $\kappa = 0.75$, from equation 7: $\theta = 40\,\text{degr}$, from equation 1: $\sigma = 46.94\,\text{degr}$, from equation 6: $\lambda = 0.8256$ and from equation 2: $\omega = 8.353\,\text{degr}$. From these data and the conditions of rule 2 stated above, it is obvious that the spiral system does not have contra-rotating branches and therefore it cannot be closed.

In order to design a closed spiral system, we specify the basic parameters φ, κ (or θ, λ), n and m and we calculate θ (or φ) from equation 5 or 7 (co-rotating or contra-rotating), λ (or κ) from equation 6, σ from equation 1 and ω from equation 2. The parameters to be specified can be considered as design parameters of the closed spiral system we want to produce. The vertex

78

numbering system in the present work covers all cases of spiral systems, closed or open, with divergence angle (section 6) or not.

In Sushida, Hizume, Yamagishi [2, p.2] we notice that the parameters n and m which have a indirect but similar effect on the shape of closed spirals, have to be relatively prime integers (something necessary only when a divergence angle has to be calculated, as shown in section 6), without any geometrical and visual meaning, thus restricting the ability to understand in a realistic way the produced spirals. We also notice that no conditions similar to those of rule 2 (and the rest of the rules which appear in the present work with their geometrical nature) exist.

In [5] a number of examples using the exponential definition of spiral systems is presented and it is quite interesting to see that the governing equations in the complex plane do not have the direct geometrical and visual meaning of such Euclidean methods as the present work.

It is interesting to note that when m = n, a condition which creates a symmetrical closed system, we get from equation 5 that: $\varphi = \theta - 2\pi/n$. This gives us: $\varphi < \theta$ for symmetrical quadrangular or triangular co-rotating closed spiral systems

The parameters n, m, φ, θ, κ, λ of figures 2b, 2c and 2d have values which appear at Table 3 of Appendix 3.

Closed Triangular Spiral Systems Where $\omega = \Theta$

As stated previously, any quadrangular spiral system can become triangular, so a closed triangular system can exist in two cases, first when $\omega = \varphi$ (figure 3a) and second when $\omega = 0$ (figure 4a). The parameters n, m, φ, θ, κ, λ of figures 3a, 3e and 4a have values which appear at Table 3 of Appendix 3. In the present section we examine the first case and calculate the relevant necessary conditions, such as the ones which hold for figure 3a, where n = 12 and m = 2.

From equation 2 and since $\omega = \varphi$, we have:

$\tan(\varphi + \sigma) = \lambda\sin\theta/(1 - \lambda\cos\theta) =>$

$\lambda = (\sin\varphi + \tan\sigma\cos\varphi) / (\sin(\theta+\varphi) + \tan\sigma\cos(\theta+\varphi))$

By using equation 1 in the above equation we get:

$\lambda = (\sin\varphi(1 - \kappa\cos\varphi) + \kappa\sin\varphi\cos\varphi) / (\sin(\theta+\varphi)(1 - \kappa\cos\varphi) + \kappa\sin\varphi\cos(\theta+\varphi)) =>$

(8) $\lambda = \sin\varphi/(\sin(\theta+\varphi) - \kappa\sin\theta)$

All the parameters of a closed spiral system are φ, θ, κ, λ, n, m, σ and ω. In order to design a triangular spiral system, we specify φ, n, m, we get θ from equation 5, or 7 (co-rotating or contra-rotating branches) and find a pair of values κ, λ which satisfy equations 8 and 6, using tools such as GraphSketch or Desmos ($\omega = \varphi$ and σ is calculated by equation 2). Given the values of φ, n, m, it is not always possible to find solutions for the values of κ, λ. Considering that the parameters κ and λ have to have values less than 1 in order to match the equations 6 and 8, we have to identify two cases:

Figure 3a

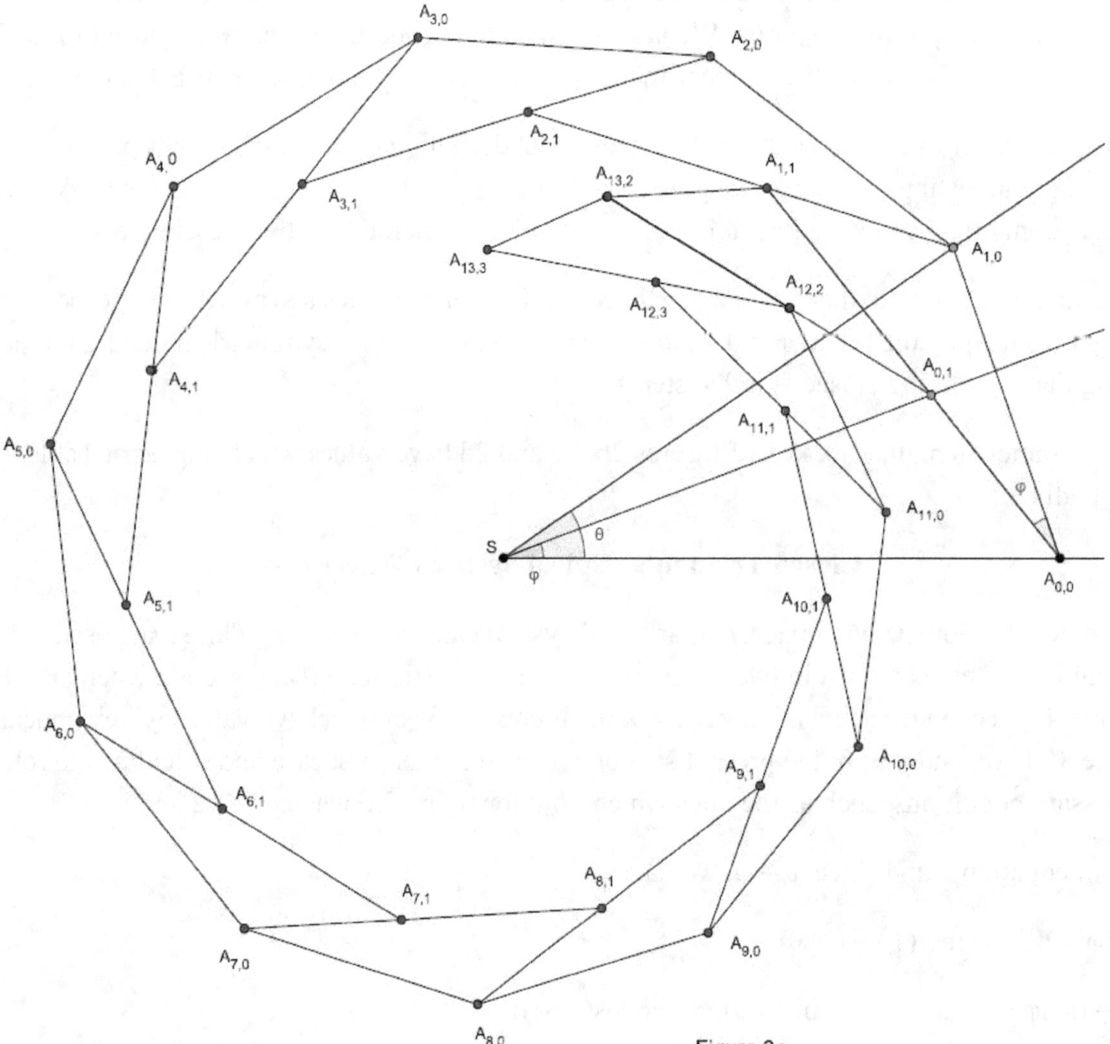

Figure 3a

Case 1: Branches $B\kappa_i$ and $B\lambda_i$ co-rotate

The parameters n, m, φ, θ are related to equation 5. Given that the branches co-rotate anticlockwise (figure 3a), equation 8 gives values of λ less than 1 as it is ($SA_{0,0} > SA_{1,0}$). In figure 3b we have a graph of equation 8 with all the important points of κ axis and λ axis. Obviously the equation 6 has its graph with concave up when n < m and concave down when n > m. By studying the range of values of these points of equation 8 in relation to equation 6, we

80

can identify the range of values of φ, n, m which allow us to calculate solutions for the values of κ, λ. table 1 gives these points and their definitions.

Figure 3b

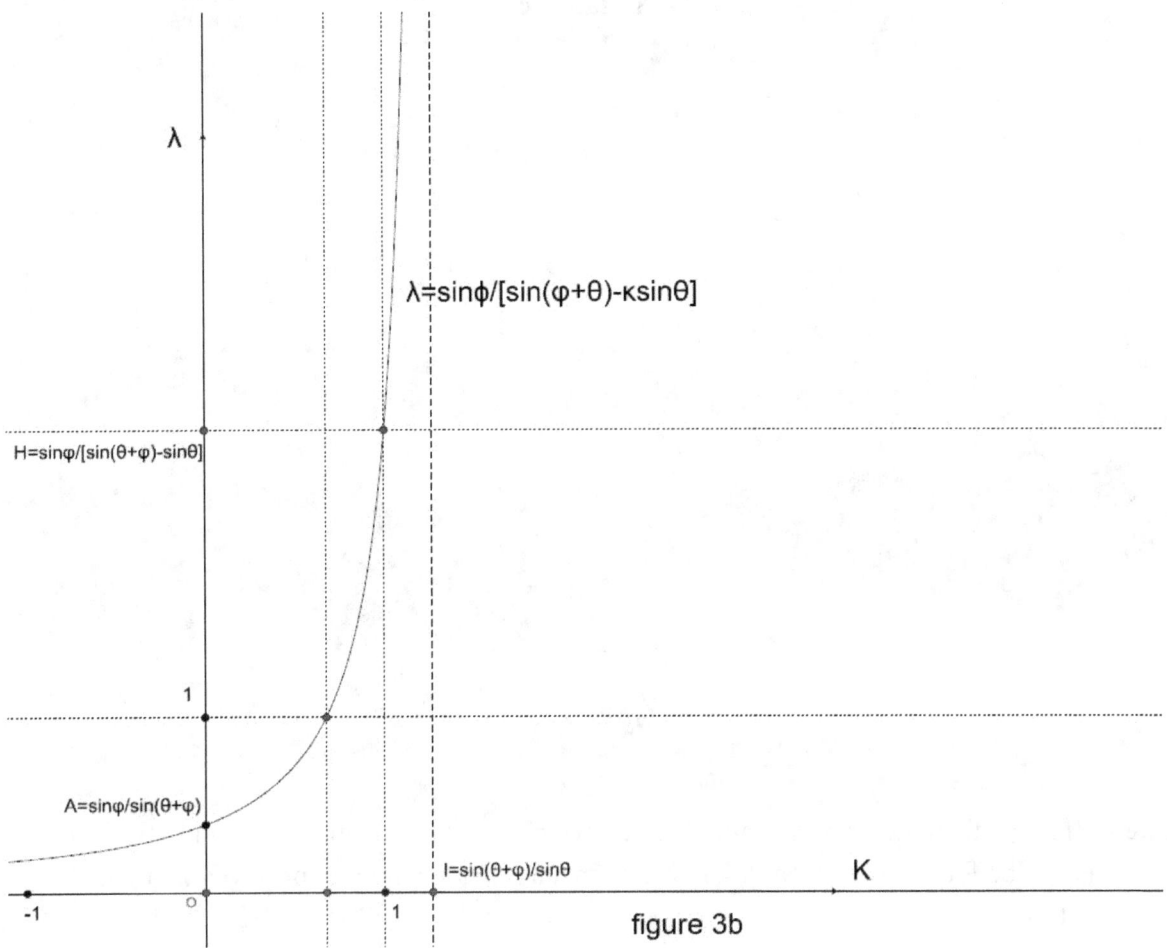

figure 3b

Table 1

$A = \sin\varphi \;/\; \sin(\theta + \varphi)$
$I = \sin(\theta + \varphi) \;/\; \sin\theta$
$H = \sin\varphi \;/\; (\sin(\theta + \varphi) - \sin\theta)$

In figure 3c we have the graphs of equations 6 and 10 for the closed co-rotating triangular spiral system of figure 3a, where we notice that there are two solutions of κ and λ values satisfying the equations, one with too small κ value for any visual presentation and the other, related with the system of figure 3a.

81

From figure 3b it is obvious that if $A \geq 1$ then there is no solution. This and some other conditions create the following rules which apply to the case of closed triangular spiral system where $\omega = \varphi$.

Figure 3c

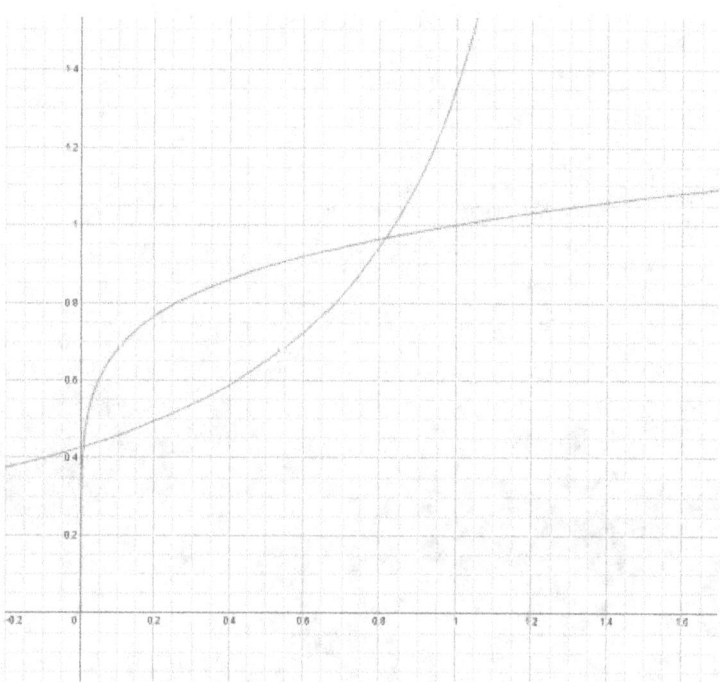

Rule 4: If $A \geq 1$ then there are no values of κ, λ which satisfy the equations 6 and 10. Therefore solutions can be found only when $A < 1$, which means $\sin\varphi < \sin(\theta+\varphi)$ or $\theta+\varphi < \pi-\varphi$ which is equivalent to:

(9) $\qquad\qquad\qquad \theta + 2\varphi < \pi$

Two immediate corollaries of this: there can be no solution when $\varphi = \theta = \pi/3$ (when the spiral triangles become equilateral) and also:

(10) $\quad \varphi < \pi/2$

Rule 5: If $I \geq 1$ then $\sin\theta < \sin(\theta+\varphi)$ or $\theta+\varphi < \pi-\theta$ which is equivalent to:

(11) $\quad 2\theta + \varphi < \pi$

This inequality together with inequality 9 give us:

(12) $\quad \theta + \varphi < 2\pi/3$

82

Rule 6: If $I < 1$ then $\sin\theta > \sin(\theta + \varphi)$ or $\theta+\varphi > \pi - \theta$ which is equivalent to:

(13) $2\theta + \varphi > \pi$

Figure 3d

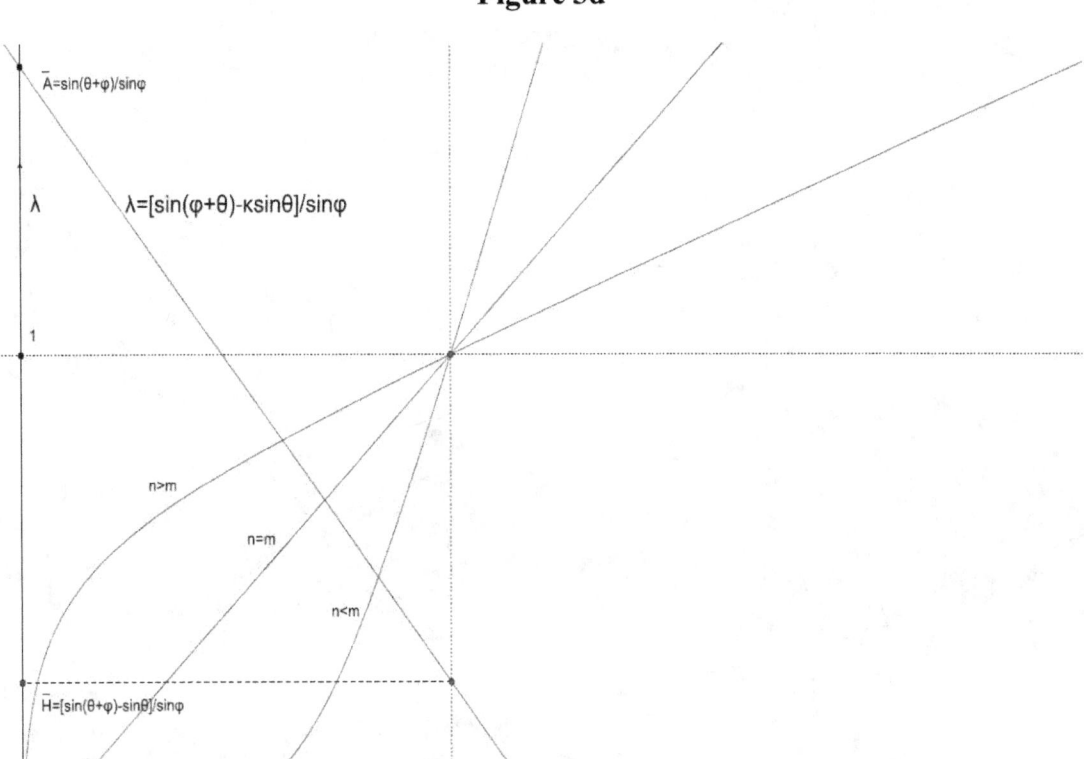

figure 3d

Rule 6a: Inequalities 9 and 13 give us the following:

(14) $$\theta - \varphi > 0$$

(15) $$(\pi - \varphi)/2 < \theta < \pi - 2\varphi$$

Rule 6b: Since $\varphi_{max} = \theta$, from inequality 15 we get:

(16) $0 < \varphi < \pi/3$

Rule 7: As proved in Appendix 2, part 1, we always have $H > 1$ when $I \geq 1$ and these conditions obviously give us two solutions, provided that Rule 4 holds and the appropriate values of φ, n, and m are chosen, such as those in figure 3c, mentioned above. When $I < 1$, it is obvious that again we have two solutions, provided that the above mentioned conditions apply. In Appendix 2, part 2 these conditions are examined and also in Table 4, maximum values of φ (which give us

83

one double solution as defined in Appendix 2 part 2) and their related κ values for specific values of n and m are presented.

Case 2: Branches $B\kappa_i$ and $B\lambda_i$ contra-rotate

Figure 3e

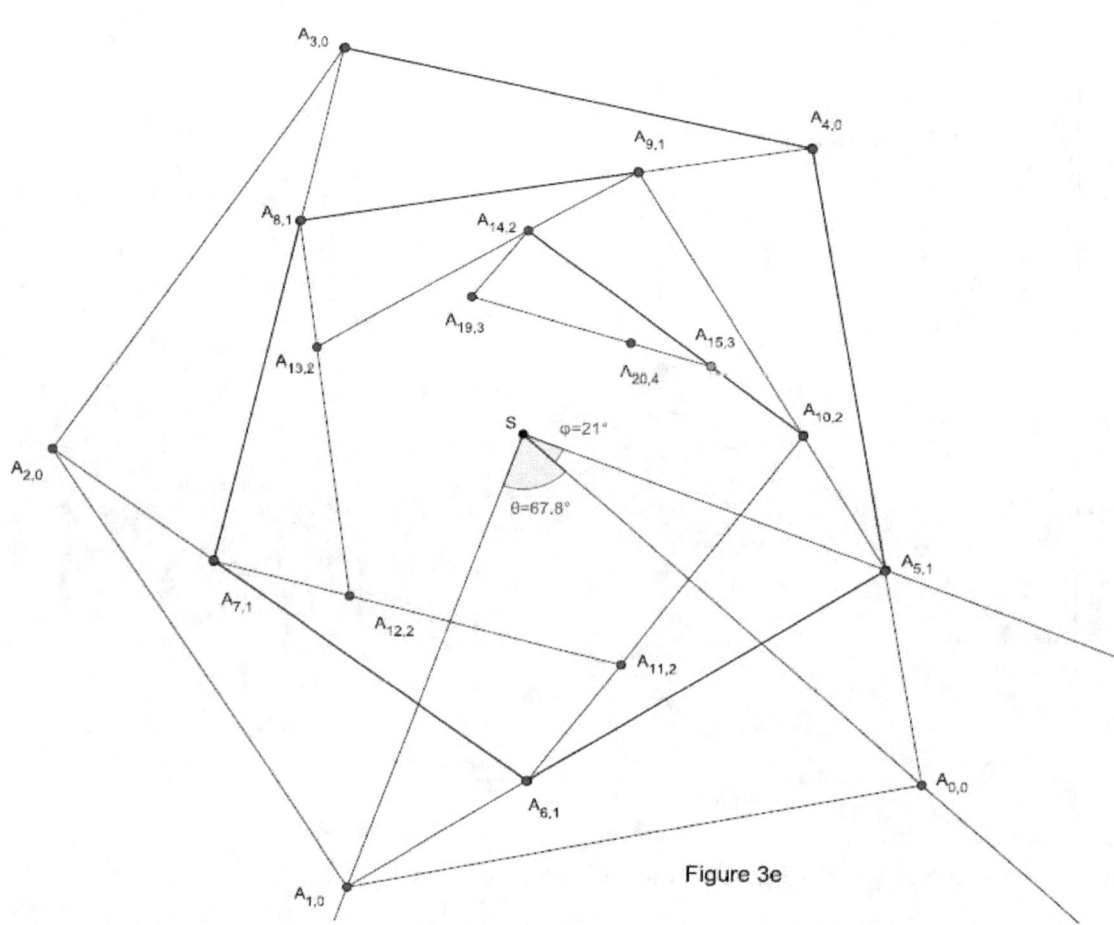

Figure 3e

The parameters n, m, φ, θ are related with equation 7. Given that the branches contra-rotate, equation 8 gives values of λ greater than 1 (because $SA_{0,0} < SA_{1,0}$, whereas $SA_{0,0} > SA_{0,1}$ and these two conditions guarantee the contra-rotating characteristics, being equivalent to rule 2 when ω = φ) so in order to match the equation 6, it has to be raised to the power of -1 and become reciprocal as follows:

$$\lambda = (\sin(\theta+\varphi) - \kappa\sin\theta)/\sin\varphi.$$

The equivalent variables *A, H, I* of table 1, become reciprocal in this case as demonstrated in table 1A.

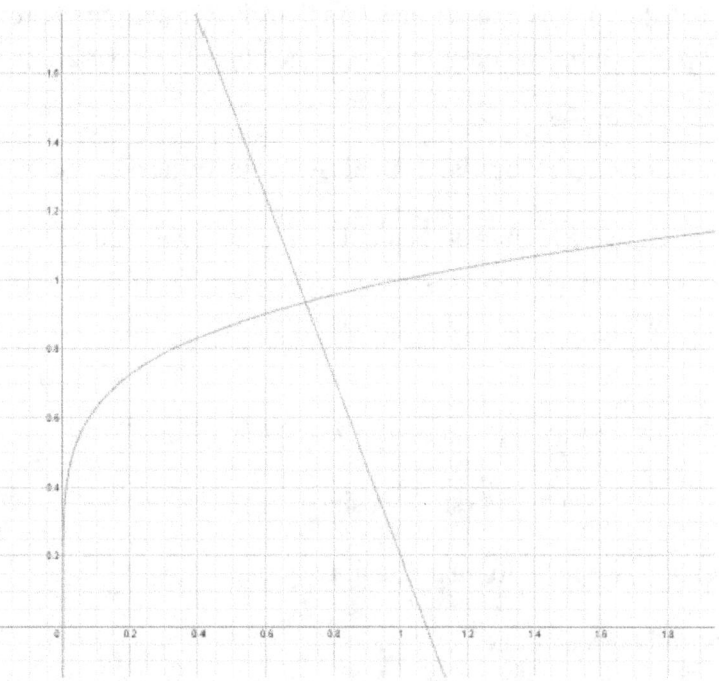

Figure 3f

Table 1a

$A' = \sin(\theta + \varphi) / \sin\varphi$
$I' = \sin\theta / \sin(\theta + \varphi)$
$H' = (\sin(\theta + \varphi) - \sin\theta) / \sin\varphi$

 In figure 3d we have a graph of the reciprocal of equation 8 which is a straight line, with all the important points of Table 1A, plus the three versions of equation 6, concave up when n < m, concave down when n > m and straight line when n = m.

Rule 8: Since we have $H > 1$ as proven in Appendix 2, part 1, it is obvious that $H' < 1$ in this case. This condition guarantees always the existence of one solution for all cases of equation 6, n < m, n = m, n > m, as shown in figure 3d. An example of such a spiral system with contra-rotating branches exists in figure 3e (the values of its parameters appear in Table 3) and the graphs of its relevant equations (the equivalent of those in graph 3d) exist in figure 3f.

85

Contrary to the case of closed quadrangular systems with contra-rotating branches, the equation 7 is sufficient together with the reciprocal version of equation 8 which takes into account the necessary condition in order to have values of λ less than 1, as mentioned previously at the beginning of CASE 2 paragraph. The only limits for the angles are : $\varphi < 2\pi/m$ and $\theta < 2\pi/m$ from equation 7.

Closed Triangular Spiral Systems Where $\omega = 0$

As in the previous section, we examine and calculate the necessary conditions to have a closed triangular spiral system in this case, such as the one of figure 4a. The parameters n, m, φ, θ, κ, λ of figure 4a have values which appear at Table 3 of Appendix 3. From equations 1 and 2 and for $\omega = 0$ we have:

$$\kappa\sin\varphi / (1 - \kappa\cos\varphi) = \lambda\sin\theta / (1 - \lambda\cos\theta)$$

(17) $$\lambda = \kappa\sin\varphi / (\sin\theta + \kappa\sin(\varphi - \theta)).$$

As in the previous section, we have to distinguish two cases:

Case 1: Branches $B\kappa_i$ and $B\lambda_i$ co-rotate

The parameters n, m, φ, θ are related with equation 5. In figure 4b, we have a graph of equation 17 with all the important points of κ and λ axis. Table 2 below gives these points and their definitions.

Table 2

$I = - \sin\theta / \sin(\varphi - \theta)$
$H = \sin\varphi / (\sin(\varphi - \theta) + \sin\theta)$

As in the previous section, by studying the range of values of these points of equation 17 in relation to equation 6, we can identify the conditions related to values of φ, n, m which allow us to calculate solutions for the values of κ, λ. These conditions create the following rules which apply to the case of closed triangular spiral system where $\omega = 0$.

Rule 9: As proved in Appendix 2, part 3, when $0 < \varphi < \theta$ we have $H > 1$ and these conditions always give us two solutions as long as n > m, one of which (for $\kappa = 0$) is without geometrical significance, as shown in figure 4c in which the graphs of equations 6 and 17 are presented for the case of figure 4a.

86

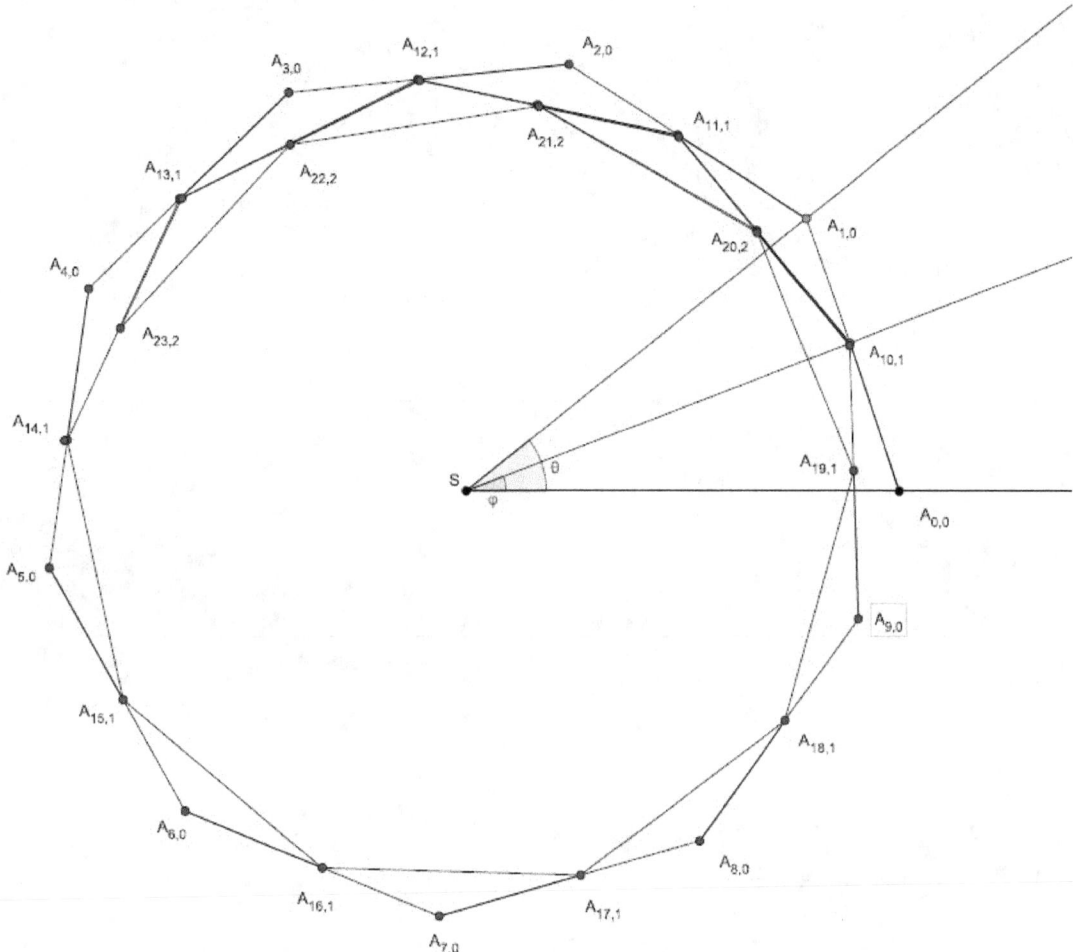

Rule 9a: When $0 < \varphi < \theta$ and if $I > 1$, then $\sin\theta > \sin(\theta - \varphi)$ or $\theta - \varphi < \pi - \theta$, which is equivalent to:

(18) $2\theta - \varphi < \pi$.

Rule 9b: When $0 < \varphi < \theta$ and if $I < 1$, then:

(19) $2\theta - \varphi > \pi$.

Rule 10: When $0 < \varphi = \theta$ then from equation 17 we get $\lambda = \kappa$, which does not produce any meaningful solution.

Rule 11: When $\varphi > \theta$ it is obvious from the definition of I that always $I < 0$.

Rule 11a: When $\varphi > \theta$ and $n < m$ it is obvious that we have two solutions, one of which (for $\kappa = 0$) is without geometrical significance.

Figure 4b

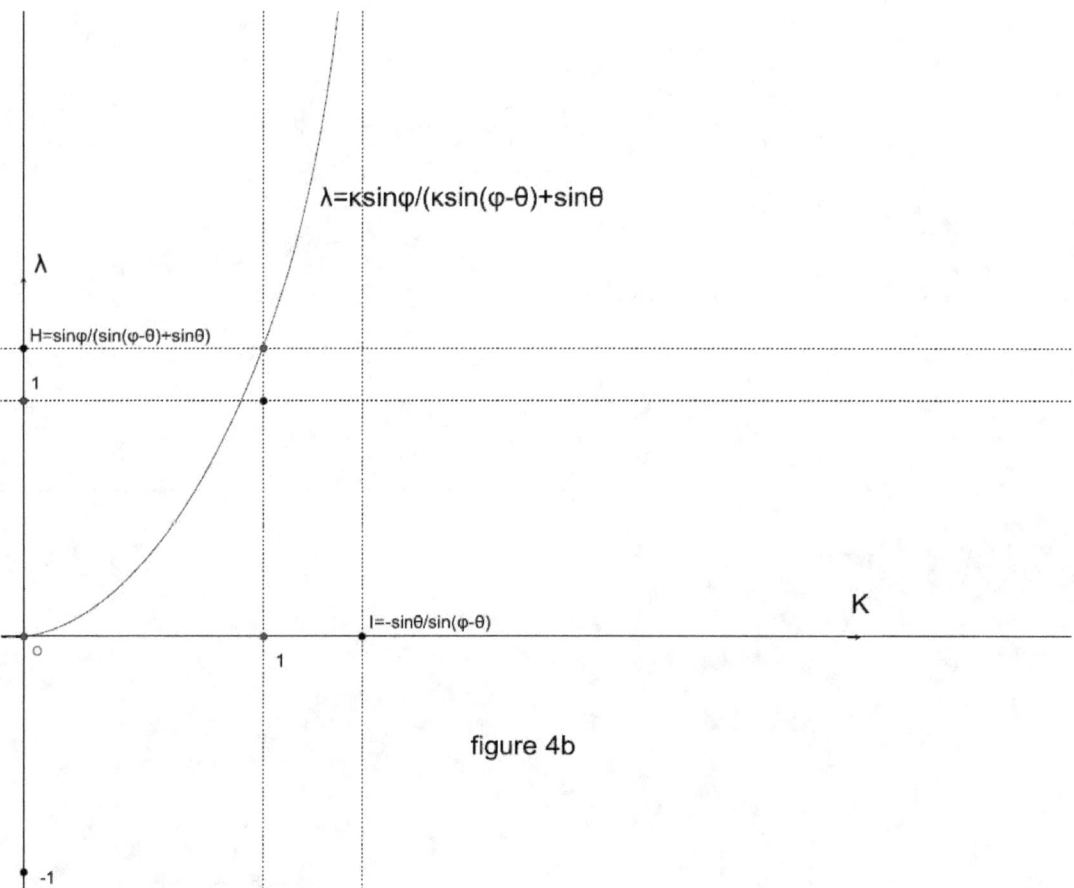

figure 4b

Case 2: Branches $B\kappa_i$ and $B\lambda_i$ contra-rotate.

The parameters n, m, φ, θ are related with equation 7. Given that the branches contra-rotate, equation 17 gives values of λ greater than 1 (because $SA_{0,0} < SA_{1,0}$, whereas $SA_{0,0} > SA_{0,1}$ as in case 2 of section 3), so in order to match the equation 6, it has to be raised to the power of -1 and become reciprocal as follows:

$\lambda = (\sin\theta + \kappa\sin(\varphi - \theta)) / \kappa\sin\varphi$.

The equivalent variable H of Table 2, becomes reciprocal in this case as follows:

$H' = (\sin(\varphi - \theta) + \sin\theta) / \sin\varphi$.

In figure 4c we have a graph of the reciprocal of equation 17 with the important point H', plus the three versions of equation 6, concave up when n < m, concave down when n > m and straight line when n = m.

Figure 4c

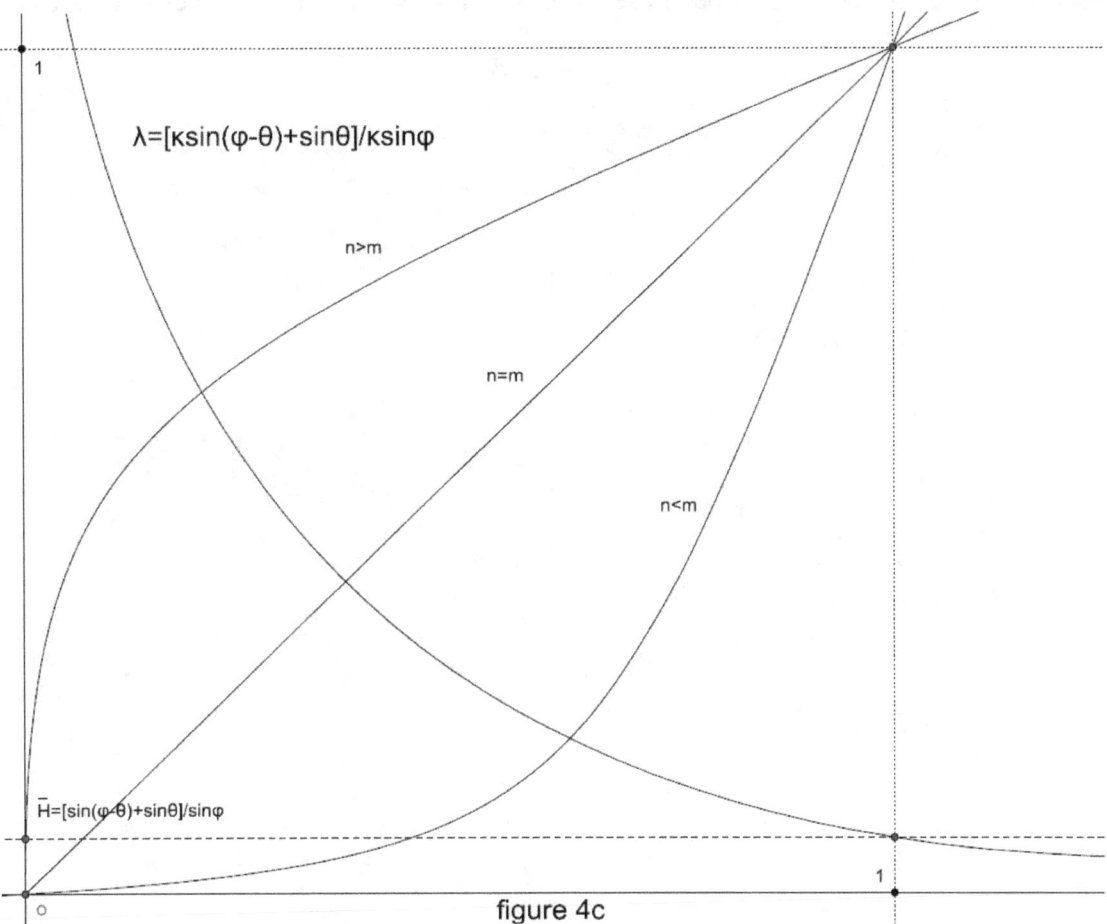

$\lambda = [\kappa\sin(\varphi-\theta)+\sin\theta]/\kappa\sin\varphi$

n>m

n=m

n<m

$\bar{H}=[\sin(\varphi-\theta)+\sin\theta]/\sin\varphi$

figure 4c

Rule 12: As proved in Appendix 2, part 3, when $0 < \varphi < \theta$ we have $H > 1$, so $H' < 1$, which gives us always one solution for all cases of equation 6, as shown in figure 4c. When $\varphi > \theta$ we have $H' > 1$, which gives us no solution.

As in section 3 and contrary to the case of closed quadrangular systems with contra-rotating branches, the equation 7 is sufficient together with the reciprocal version of equation 17 which takes into account the necessary condition in order to have values of λ less than 1, as mentioned previously at the beginning of case 2 paragraph.

Equivalence of Closed Triangular Spiral Systems Where Either $\omega = 0$ or $\omega' = \Phi$

As stated in the introduction, we can treat the case of a spiral system where $\omega = 0$ just like the case of a new spiral system where $\omega' = \varphi$ by introducing the parameters λ', m', and the angle θ' of the new system, which has the other parameters (κ, φ and n) the same with the first one. In this

89

section we will examine the necessary relations in order to have the equivalence of the two types of closed triangular systems, described in sections 3 and 4. We have all together five cases of

Figure 4d

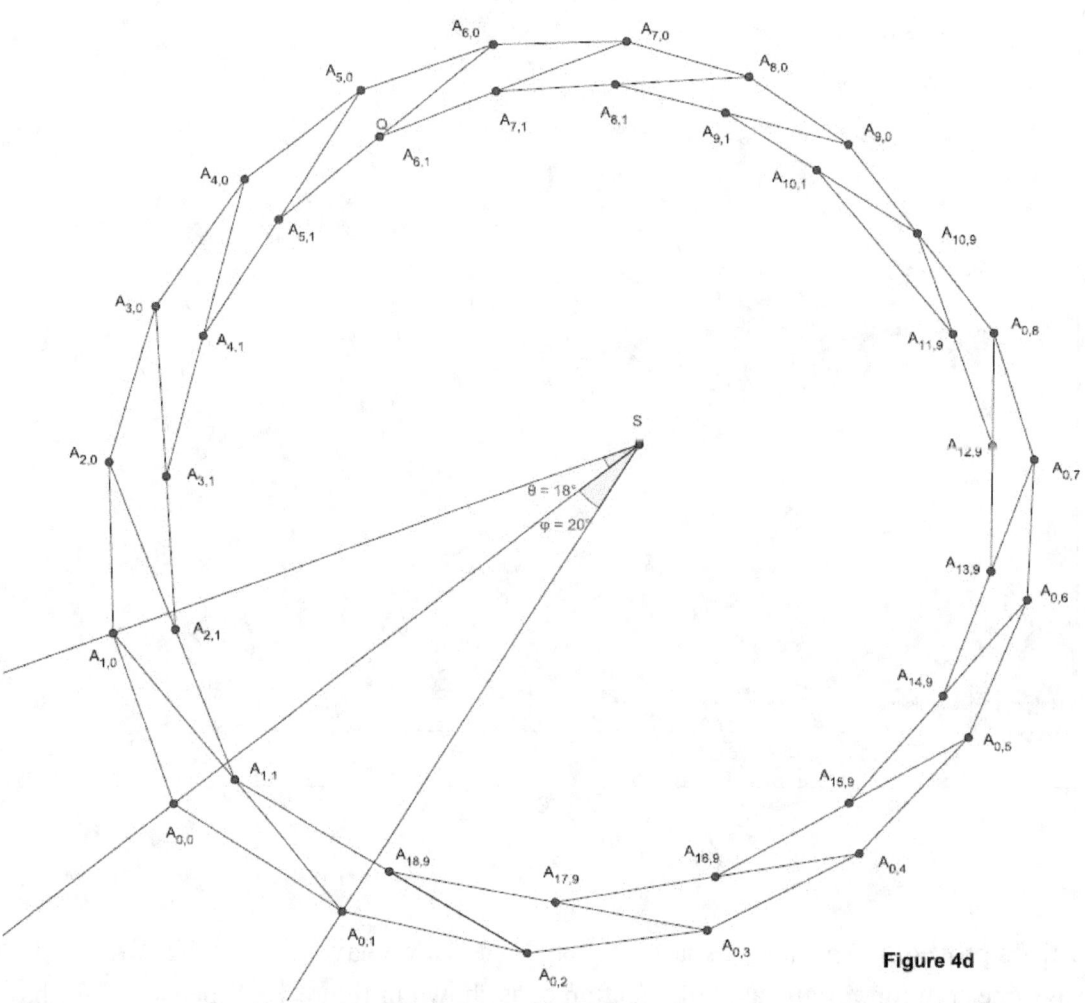

Figure 4d

equivalence, presented below. It has to be noted that the analysis and the equations 20 to 32 can be applicable also to any quadrangular and triangular spirals.

Case 1: Branches $B\kappa_i$ and $B\lambda_i$ co-rotate, $\theta > \varphi$ and $n > m$.

In this case we have equation 5 and equation 6 for the old system as shown in figure 4a, where we have $\lambda' = A_{20,2}A_{11,1}/A_{11,1}A_{2,0} = \kappa/\lambda$ and $\theta' = {}^\wedge A_{20,2}SA_{11,1} = {}^\wedge A_{11,1}SA_{2,0} = \theta - \varphi$ similarly to figure 1d, taking in consideration that the branches $B\kappa_i$ and $B\lambda_i$ co-rotate, $m < n$ and $\kappa < \lambda$.

In Appendix 4 we calculate equation 20 and it is obvious that the relevant equations of the other cases can be calculated in a similar way. In figure 4d we have the equivalent new system ($\omega' = \varphi$)

to the old system ($\omega=0$) related to figure 4a. The parameters n, m$'$, φ, θ', κ, λ' of figure 4d have values which appear at Table 3 of Appendix 3. The following equations apply for both systems:

(20) $\qquad\qquad\qquad n\theta' = 2\pi - m'\varphi = 2\pi - (n-m)\varphi$

(21) $\qquad\qquad\qquad\qquad \lambda' = \kappa / \lambda$

(22) $\qquad\qquad\qquad \lambda' = \sqrt[n]{\kappa^{m'}} = \sqrt[n]{\kappa^{n-m}} = \kappa^{(n-m)/n}.$

At the equation 20 we have that $\theta' = \theta - \varphi$ and $m' = n - m$, therefore equations 5 and 20 are equivalent, also the branches of the new system contra-rotate. By supposing that $\kappa < 1$ and because $n > m$ and equations 6 and 21 hold, we have that $\lambda' < 1$ and $\kappa < \lambda$. Finally from equations 6, 21 and 22 we have:

$$\lambda' = \kappa / \lambda = \kappa^{(n-m)/n} \quad \text{and} \quad \lambda = \sqrt[n]{\kappa^m} = \kappa^{m/n} \Rightarrow \kappa = \kappa^{m/n} \, \kappa^{(n-m)/n} = \kappa.$$

From the above we deduce that all the equations of the old and the new system are equivalent (being of the same type) and compatible.

Case 2: Branches $B\kappa_i$ and $B\lambda_i$ co-rotate, $\theta > \varphi$ and $n < m$.

In this case we have equation 5 and equation 6 for the old system, plus the following equations for the new system:

(23) $\qquad\qquad\qquad n\theta' = 2\pi + m'\varphi = 2\pi + (m-n)\varphi$

(24) $\qquad\qquad\qquad\qquad \lambda' = \lambda / \kappa$

(25) $\qquad\qquad\qquad \lambda' = \sqrt[n]{\kappa^{m'}} = \sqrt[n]{\kappa^{m-n}} = \kappa^{(m-n)/n}.$

The branches of the new system co-rotate. By supposing that $\kappa < 1$ and because $n < m$ and equations 6 and 24 hold, we have that $\lambda' < 1$ and $\kappa > \lambda$. From equations 6, 24 and 25 we have:

$$\lambda' = \lambda / \kappa = \kappa^{(m-n)/n} \quad \text{and} \quad \lambda = \sqrt[n]{\kappa^m} = \kappa^{m/n} \Rightarrow 1 / \kappa = \kappa^{-m/n}\kappa^{(m-n)/n} = 1 / \kappa$$

Case 3: Branches $B\kappa_i$ and $B\lambda_i$ co-rotate, $\theta < \varphi$ and $n < m$.

In this case we have the following equation (a version of equation 5) and equation 6 for the old system:

(26) $\quad m\varphi = 2\pi + n\theta.$

We also have the following equations for the new system:

(27) $$n\theta' = 2\pi - m'\varphi = 2\pi - (m-n)\varphi$$

(28) $$\lambda' = \lambda / \kappa$$

(29) $$\lambda' = \sqrt[n]{\kappa^{m'}} = \sqrt[n]{\kappa^{m-n}} = \kappa^{(m-n)/n}.$$

The branches of the new system contra-rotate. By supposing that $\kappa < 1$ and because $n < m$ and equations 6 and 28 hold, we have that $\lambda' < 1$ and $\kappa > \lambda$. Finally from equations 6, 28 and 29 we have:

$$\lambda' = \lambda / \kappa = \kappa^{(m-n)/n} \quad \text{and} \quad \lambda = \sqrt[n]{\kappa^m} = \kappa^{m/n} \Rightarrow 1/\kappa = \kappa^{-m/n}\kappa^{(m-n)/n} = 1/\kappa$$

Case 4: Branches $B\kappa_i$ and $B\lambda_i$ co-rotate, $\theta < \varphi$ and $n > m$.

In this case we have equation 26 and equation 6 for the old system, plus the following equations for the new system:

(30) $$n\theta' = 2\pi + m'\varphi = 2\pi + (n-m)\varphi$$

(31) $$\lambda' = \kappa / \lambda$$

(32) $$\lambda' = \sqrt[n]{\kappa^{m'}} = \sqrt[n]{\kappa^{n-m}} = \kappa^{(n-m)/n}.$$

The branches of the new system co-rotate. By supposing that $\kappa < 1$ and because $n > m$ and equations 6 and 31 hold, we have that $\lambda' < 1$ and $\kappa < \lambda$. From equations 6, 31 and 32 we have:

$$\lambda' = \kappa / \lambda = \kappa^{(n-m)/n} \quad \text{and} \quad \lambda = \sqrt[n]{\kappa^m} = \kappa^{m/n} \Rightarrow \kappa = \kappa^{m/n}\kappa^{(n-m)/n} = \kappa.$$

Case 5: Branches $B\kappa_i$ and $B\lambda_i$ contra-rotate.

In this case we have equation 7 and equation 6 for the old system, plus the following equations (equation 34 is calculated at Appendix 1.5) for the new system:

(33) $$n\theta' = 2\pi - m'\varphi = 2\pi - (n+m)\varphi$$

(34) $$\lambda' = \kappa\lambda$$

(35) $$\lambda' = \sqrt[n]{\kappa^{m'}} = \sqrt[n]{\kappa^{n+m}} = \kappa^{(n+m)/n}.$$

The branches of the new system contra-rotate. By supposing that $\kappa < 1$

and $\lambda < 1$, we have that $\lambda' < 1$. From equations 6, 34 and 35 we have:

$$\lambda' = \kappa\lambda = \kappa^{(n+m)/n} \quad \text{and} \quad \lambda = \sqrt[n]{\kappa^m} = \kappa^{m/n} \Rightarrow \kappa = \kappa^{-m/n}\kappa^{(n+m)/n} = \kappa.$$

The parameters listed below belong to an old and a new equivalent spiral system related to this case, following the conditions of rule 2: m = 2, n = 8, φ = 15degr, θ = 41.25degr, κ = 0.945, λ = 0.986, σ = 70.38degr, λ′= 0.932, θ = 26.25degr, m′= 10.

Figure 5

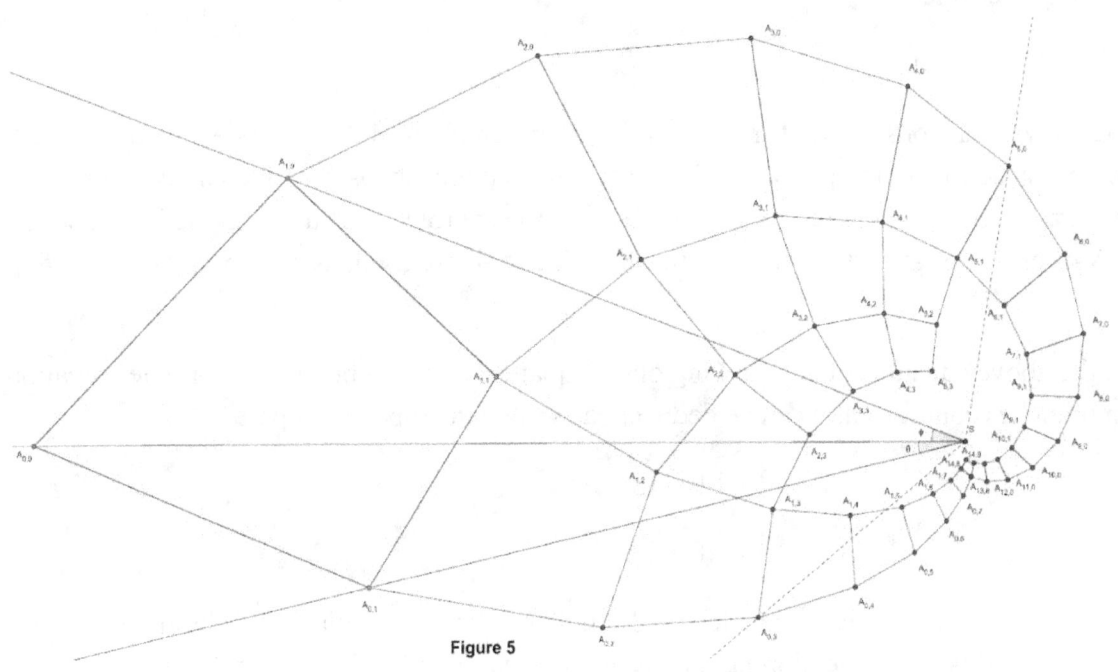

Figure 5

Divergence Angle

The divergence angle d, as described in [2, p.2 and p.3], (figure 1, angle formed by vertex 1, centre of spiral O and vertex 2), in [3, p.100], in [4, p.20] and in figure 5 (present work) is defined by drawing lines from two successive primordia to the center of the spiral and d > π or d < π according to the direction of rotation the angle is measured. As it is stated in [4, p.19] (Bravais-Bravais theorem), given any starting vertex (primordium) at a spiral system with parastichies (m,n), the next vertex found in the m parastichy passing from the starting vertex (in the λ direction, according to the present work) differs by m (from the starting vertex) and similarly the next vertex found in the n parastichy passing from the starting vertex (in the κ direction) differs by n (from the starting vertex, example in [2, p2]). This means that starting from any vertex of the spiral tiling-system, one has to turn md times to go to the next vertex of the m parastichy passing from the starting point and nd times to go to the next vertex of the n parastichy passing from the starting point. This rule can be expressed by the following obvious equations, applicable in the spiral systems described in the present work, when the branches of κ and λ directions (n and m parastichies according to [4]) co-rotate:

93

(36) $2\pi I_\lambda - \varphi = nd$

(37) $2\pi I_\kappa - \theta = md.$

Similarly, when the branches of κ and λ directions contra-rotate, equation 37 remains and equation 36 becomes:

(38) $2\pi I_\lambda + \varphi = nd.$

In the above equations, I_λ and I_κ represent the integer number of 2π or 360degr. turns which are required for the rotation to go to the next vertex of each branch, as described above (for examples, see figure 5 and further below). If the opposite rotation is to be applied, the above equations change sign and different values I'_λ, I'_κ and d' are used, as shown in Appendix 5, part 2.

From the above equations, the following ones (equation 39 when branches co-rotate, equation 40 when branches contra-rotate) can be deduced, as shown in Appendix 5 part 1:

(39) $d = I_\lambda\theta - I_\kappa\varphi$

(40) $d = I_\lambda\theta + I_\kappa\varphi.$

In Table 3 the values of the parameters I_κ, I_λ and d have been calculated for the figures of the present work. It is worth mentioning that at figure 3a where n = 12 and m = 2, therefore are non relatively prime, we have no calculation of d. Also in the example of figure 2b (m = 1, n = 13, d = 331.2degr), if the point $A_{13,1}$ is chosen as starting point, then going clockwise $I_\lambda = 12$ times with angle increments of $\theta = 28.8$degr, we arrive at point $A_{1,0}$. From there we go anticlockwise with $I_\kappa = 1$ increment of $\varphi = 14.6$degr., we arrive at point $A_{14,1}$, implementing equations 37 and 39 once (m = 1), in order to get what we stated at the beginning of this section, "one has to turn md times to go to the next vertex of the m parastichy passing from the starting point". Applying this process another 12 times, therefore a total of 13 times (n = 13), we arrive at point $A_{26,2}$, implementing equation 36, in order to have turned "nd times to go to the next vertex of the n parastichy passing from the starting point". The case of figure 5 is based on [2, p2, contra-rotating example of figure 1]. The relevant parameters are, m = 13, n = 8, $\theta = 12.5$degr, $\varphi = 20$degr, d = 137.5degr, κ = 0.78, λ = 0.668, $I_\lambda = 3$, $I_\kappa = 5$ (the last two parameters were calculated by using equation 41), which all comply with the above mentioned equations. In figure 5, assuming that the starting vertex is $A_{0,3}$, we have that $d = {}^\wedge A_{0,3}SA_{5,0} = 137.5$degr. It is interesting to notice that the ray starting from point S and passing from point $A_{0,3}$ (first primordium) goes through points $A_{0,3} \rightarrow A_{1,3} \rightarrow A_{2,3} \rightarrow A_{3,3} \rightarrow A_{4,3} \rightarrow A_{5,3}$ thus turning 5x20degr (κ direction) and continues going through points $A_{5,3} \rightarrow A_{5,2} \rightarrow A_{5,1} \rightarrow A_{5,0}$ thus turning another 3x12.5degr (λ direction), in order to pass from point $A_{5,0}$ (second primordium). This is a visual application of equation 40. The visual applications of equations 38 and 37 are produced by assuming that this ray (starting from $A_{0,3}S$) has to rotate nd (8x137.5degr) to go to $A_{1,3}S$ ($^\wedge A_{0,3}SA_{1,3} = \varphi = 20$degr, equation 38) and md (13x137.5degr) to go to $A_{0,4}S$ ($^\wedge A_{0,3}SA_{0,4} = \theta = 12.5$degr, equation 37).

The following Diophantine equation is deduced from equations 36, 37, 5 when branches co-rotate or 37, 38 and 7 when branches contra-rotate (Appendix 5, part 2, where it is also shown that the sign + becomes − when alternative versions of equations 36, 37, 38, opposite rotation, I'_λ, I'_κ and d′ are used) :

(41) $nI_\kappa - mI_\lambda = +1$.

It is important to note the following: The equations 36, 37 (branches co-rotating), or 37, 38 (branches contra-rotating) have 3 unknowns, I_κ , I_λ and d, plus the condition that I_κ and I_λ are integers. Let n = vq and m = vp (where v, p, q, integers plus p and q relatively prime integers), in other words, n and m non relatively prime integers. From the above assumption and equation 41, we get: $qI_\kappa - pI_\lambda = 1/v$. This equation is not valid, since I_κ, I_λ, p, q, v are integers. Therefore, the following is proven:

Rule 13: At any closed spiral system, if m and n are not relatively prime integers, there is no possibility for a convergence angle to exist.

Equation 41, can always be solved as a Diophantine equation, where n, and m are relatively prime integers.

Finally, according to [4, p.20], [2, p.3], the plastochrone R ratio is the ratio of distances between two successively numbered primordia and the center of the spiral and R > 1. This ratio can be defined according to the present work, using equation 39 or 40 in order to go from any primordium to the next one, after a rotation of d (see figure 5). This definition is as follows, as shown in Appendix 5, part 3:

(42) $r = (1/\lambda)^{I_\lambda}\kappa^{I_\kappa} = \kappa^{1/n} = \lambda^{1/m} < 1 \Rightarrow R = 1/r > 1$.

In figure 5, according to the definition of plastochrone ratio, $R = A_{0,3}S/A_{5,0}S$ and R = 1/r, therefore we also have a visual application of equation 42

$(r = (1 / 0.668)^3(0.78)^5 = (0.78)^{1/8} = 0.97)$.

In Appendix 5, part 4, an equivalence is shown between specific equations of the present work and of equations at [2].

All the equations produced in the present work do not have any approximation nature, since they fully comply with the used analytical and geometrical approach (in [4], some formulae which are

95

quite similar to those in the present work appear to have an approximation nature, for example in [4, p. 37 and p. 43]).

Final Remarks

The purpose of this work is to enable the analysis and design of closed spiral systems using a simple Euclidean geometry approach, offering thus a full visual understanding of the parameters which govern these systems. This approach can be enhanced by further research which can produce some practical applications in phyllotaxis problems which could be combined with theorems or methods based in the already existing complex plane theories. Last but not least, one can design as many variants of spiral systems as he can imagine by tuning the appropriate design parameters and this can give pleasure to the designer.

Appendices

Appendix 1

Part 1: In figure 1a, let the ray $A_{0,0}A_{0,1}$ pass from a random point C not belonging to the segment $A_{0,0}A_{0,1}$ and similarly the ray $SA_{0,1}$ intersect the seg-ment $A_{0,0}A_{1,0}$ at point B. Let $\varphi' = {^\wedge}A_{0,2}A_{0,1}C$. Since the triangles $\triangle A_{0,0}A_{0,1}S$ and $\triangle A_{0,1}A_{0,2}S$ are similar and due to the fact that ${^\wedge}SA_{0,1}A_{0,2} + {^\wedge}A_{0,2}A_{01}C + {^\wedge}CA_{0,1}B = \pi$, as a result $\varphi' = \varphi$ and ${^\wedge}A_{0,0}A_{0,1}A_{0,2} = \pi - \varphi$, so:

$(A_{0,0}A_{0,1}{^\wedge}A_{0,1}A_{0,2}) = \varphi$.

Part 2: By applying the sin law in the similar triangles $\triangle A_{0,0}A_{0,1}S$ and

$\triangle A_{0,1}A_{0,2}S$ we get: $SA_{0,0}/\sin({^\wedge}A_{0,0}A_{0,1}S) = SA_{0,1}/\sin\varsigma$.

Therefore $\sin({^\wedge}A_{0,0}A_{0,1}S) = \sin\sigma/\kappa$ since $\kappa = SA_{0,1}/SA_{0,0}$. Taking into account that $\pi - ({^\wedge}A_{0,0}A_{0,1}S) = \varphi + \sigma$, so $\sin({^\wedge}A_{0,0}A_{0,1}S) = \sin\varphi\cos\sigma + \cos\varphi\sin\sigma$, we deduce from the above:

$\tan\sigma = \kappa\sin\varphi/(1 - \kappa\cos\varphi)$.

Part 3: From figure 1a, we get

$(A_{0,0}A_{0,1}{^\wedge} A_{1,0}A_{1,1}) = {^\wedge}A_{0,0}DA_{1,0}$ also ${^\wedge}A_{0,0}A_{1,0}D = {^\wedge}A_{0,0}A_{1,0}S + {^\wedge}SA_{1,0}D$

and ${^\wedge}A_{1,0}A_{0,0}D = \omega$. From $\triangle A_{0,0}A_{1,0}D$ we get ${^\wedge}A_{0,0}DA_{1,0} = \pi - {^\wedge}A_{1,0}A_{0,0}D - {^\wedge}A_{0,0}A_{1,0}D$. From $\triangle A_{0,0}A_{1,0}S$ we get ${^\wedge}A_{0,0}A_{1,0}S = \pi - \sigma - \omega - \theta$ and from segment $A_{1,0}A_{1,1}$ of branch $B_{\kappa1}$ we get ${^\wedge}SA_{1,0}A_{1,1} = \sigma$, therefore:

${^\wedge}A_{0,0}A_{1,0}D = \pi - \theta - \omega$.

From all the above, we get:

$(A_{0,0}A_{0,1}{}^{\wedge}A_{1,0}A_{1,1}) = \theta.$

Part 4: From figure 1a and because $^{\wedge}A_2,A_{1,0}A_{1,1} = \omega$ we get that:

$^{\wedge}A_{0,0}A_{1,0}A_{1,1} = \pi - \theta - \omega$ and $^{\wedge}A_{0,0}A_{1,0}S = \pi - \theta - \omega - \sigma$

In the same figure, since $^{\wedge}A_{1,1}A_{0,1}A_{0,2} = \omega$, we get that:

$^{\wedge}A_{0,0}A_{0,1}A_{1,1} = \pi + \varphi - \omega$

In the same figure, we also have that $^{\wedge}A_{0,1}A_{1,1}A_{1,0} = \pi - \varphi - {}^{\wedge}A_{0,1}A_{1,1}A_{1,2}$ and because $^{\wedge}A_{0,1}A_{1,1}A_{1,2} = \pi - \theta - \omega$, this gives us:

$^{\wedge}A_{0,1}A_{1,1}A_{1,0} = \theta + \omega - \varphi.$

In figure 1c where $\omega = \varphi$, the above relations become:

$^{\wedge}A_{0,0}A_{1,0}A_{1,1} = \pi - \theta - \varphi$ plus $^{\wedge}A_{0,0}A_{0,1}A_{1,1} = \pi$ and $^{\wedge}A_{0,1}A_{1,1}A_{1,0} = \theta.$

In figure 1d where $\omega = 0$, the above relations become:

$^{\wedge}A_{0,0}A_{1,0}A_{1,1} = \pi - \theta$ plus $^{\wedge}A_{0,0}A_{0,1}A_{1,1} = \pi + \varphi$ and $^{\wedge}A_{0,1}A_{1,1}A_{1,0} = \theta - \varphi.$

Part 5: From figure 1d (applicable also to figures 1a, 1b, 1c) we have: $\lambda = A_{1,1}A_{2,0}/A_{0,1}A_{1,0}$ and $\kappa = A_{0,2}A_{1,1}/A_{0,1}A_{1,0}$, as ratios of diagonals to the relevant similar quadrangles (when $\omega \neq 0$) which in this case become triangles. Also from this figure we get that the branches $B\kappa_i$ and $B\lambda_i$ are co-rotating and $\kappa < \lambda$. From these two, we get:

$$\lambda' = A_{0,2}A_{1,1}/A_{1,1}A_{2,0} = \kappa/\lambda.$$

Also from the same figure, since $^{\wedge}A_{0,1}SA_{1,0} = \theta - \varphi$, due to similarity of triangles

$\Delta A_{0,1}A_{1,0}S$, $\Delta A_{0,2}A_{1,1}S$ and $\Delta A_{1,1}A_{2,0}S$, we get:

$\theta = A_{0,2}SA_{1,1} = A_{1,1}SA_{2,0} = \theta - \varphi.$

In the case where the branches $B\kappa_i$ and $B\lambda_i$ are contra-rotating, we have that $1/\lambda = A_{1,1}A_{2,0}/A_{0,1}A_{1,0}$ and $\kappa = A_{0,2}A_{1,1}/A_{0,1}A_{1,0}$, so it can be deduced from the above that:

$\lambda' = A_{0,2}A_{1,1}/A_{1,1}A_{2,0} = \kappa\lambda$

Appendix 2

Part 1: From Table 1 we have that $H = \sin\varphi/(\sin(\theta+\varphi) - \sin\theta)$, which is equation 8 for $\kappa = 1$. If we consider φ as a variable and the rest of the entities as constants with given value, then $H = f(\varphi)$, where $\varphi < \pi$. The first derivative of the above function is as follows:

$f'(\varphi) = \sin\theta(1 - \cos\varphi)/(\sin(\theta+\varphi) - \sin\theta)^2 > 0$ and $0 < \varphi < \pi$.

For $\varphi = 0$ and following Hopital's rule, we have the following:

$\lim(f(\varphi)) = [\sin\varphi]'/[(\sin(\theta+\varphi) - \sin\theta)]' = 1/\cos\theta > 1$ where $\varphi \to 0$,

therefore always $H > 1$ under the conditions of Rule 7.

Part 2: Let $A = f(\varphi) = \sin\varphi = \sin(\varphi + \theta) = \sin\varphi/\sin[(2\pi + (m+n)\varphi)/n]$, if we use equation 5. The first derivative of this function is as follows:

$A' = f'(\varphi) = \{\cos(\varphi)\sin[(2\pi + (m+n)\varphi)/n] - [(m+n)/n]\sin(\varphi)\cos[(2\pi + (m+n)\varphi)/n]\}/\{\sin[(2\pi +$

$(m+n)\varphi)/n]\}^2$.

If we assume that $A' > 0$ then this is equivalent to :

$g(\varphi) = \tan[(2\pi + (m+n)\varphi)/n] - [(m+n)/n]\tan(\varphi) > 0$,

of which if we take the first derivative, we have:

$g'(\varphi) = [(m+n)/n]\{\sec[(2\pi + (m+n)\varphi)/n]\}^2 - [(m+n)/n]\{\sec(\varphi)\}^2 > 0$

which is true since

$|\sec[(2\pi + (m+n)\varphi)/n]| > |\sec(\varphi)|$ for $\theta + \varphi < \pi - \varphi$ or $A < 1$.

So all together we have $A' > 0$ for $0 < A < 1$ or for $0 < \varphi < ((n-2)\pi) / (m+2n)$ from equation 5 and $\theta + \varphi < \pi - \varphi$. Also $A \to 0$ when $\varphi \to 0$, from equation 8 we have $\lambda \neq 0$ for $0 < \kappa < I$ and the graph of equation 8 is always concave up for $0 < \kappa < I$. Therefore for small enough values of φ there will always be pairs of values of κ and λ satisfying equations 6 and 8, such as in figure 3c. More specifically, for a given value of φ, two pairs of κ and λ exist (two solutions), apart from the case where the curve of equation 6 is tangent to the curve of equation 8 giving the maximum value of φ and its related κ value (double solution), as in figure 3c1. This is the case of the maximum value of φ to be obtained for a given pair of n and m values and in this figure we have m = 2, n = 12, maximum φ = 41.2degr and κ = 0.232. In Table 4 we have maximum values of φ and their related κ values for specific values of n and m which can be easily extended with the use of a graphic tool such as Desmos. It is useful to note that there are no closed triangle spiral cases for $0 < n < 3$, because of equation 5 and of the condition $0 < \theta < \pi$.

Part 3: From Table 2 we have that $H = \sin\varphi / (\sin(\varphi - \theta) + \sin\theta)$, which is equation 17 for $\kappa = 1$. If we consider φ as a variable and the rest of the entities as constants with given value, then $H = f(\varphi)$, where $\varphi < \pi$. The first derivative of the above function is as follows:

$f'(\varphi) = \sin\theta(\cos\varphi - 1) / (\sin(\varphi - \theta) + \sin\theta)^2 < 0$ and $0 < \varphi < \pi$.

For $\varphi = 0$ and following Hopital's rule, we have the following:

$\lim(f(\varphi)) = [\sin\varphi]' / [(\sin(\varphi - \theta) + \sin\theta)]' = 1/\cos\theta > 1$ where $\varphi \to 0$.

Additionally for $\varphi = \theta$ we have $H = f(\varphi) = f(\theta) = 1$. From all the above we can deduce that: $H > 1$ when $0 < \varphi < \theta$ and $H < 1$ when $\theta < \varphi$.

Appendix 3

TABLE 3 (angles in degrees)

figure	n	m	φ	θ	κ	λ	I_λ	I_κ	d
2b	13	1	14.60	28.80	0.484	0.946	12	1	331.2
2c	13	2	15.00	30.00	0.486	0.895	6	1	165
2d	11	1	24.00	30.54	0.512	0.941	10	1	329.46
3a	12	2	20.00	33.33	0.820	0.967	-	-	-
3e	5	1	21.00	67.80	0.718	0.936	4	1	292.2
4a	10	1	20.00	38.00	0.94	0.994	9	1	322
4d	10	9	20.00	18.00	0.94	0.945	1	1	38
5	8	13	20.00	12.50	0.78	0.668	5	3	137.5

Table 4 [φ maximum (degrees)-κ]

m-n	3	4	5	6	7	8	9
1	15.8-0.187	25.8-0.172	32.6-0.157	37.5-0.151	41.3-0.149	44.2-0.142	46.5-0.134
2	11.1-0.328	18.8-0.299	24.4-0.277	28.6-0.264	31.8-0.256	34.4-0.254	36.6-0.245
3	8.5-0.441	14.8-0.390	19.5-0.365	23.2-0.346	26.2-0.339	28.5-0.329	30.6-0.324
4	7.0-0.507	12.3-0.468	16.4-0.445	19.6-0.419	22.3-0.402	24.5-0.395	26.4-0.380
5	5.9-0.571	10.5-0.524	14.1-0.493	17.0-0.471	19.5-0.457	21.5-0.445	23.2-0.437
6	5.1-0.613	9.1-0.570	12.4-0.539	15.1-0.517	17.3-0.505	19.1-0.490	20.8-0.480
7	4.5-0.651	8.1-0.609	11.1-0.578	13.5-0.553	15.6-0.540	17.3-0.534	18.8-0.518
8	4.0-0.681	7.3-0.642	10.0-0.611	12.2-0.588	14.2-0.569	15.8-0.558	17.2-0.550

Appendix 4

Let equation 5 to be related with the old system and the equivalent equation of the new system (of which the branches contra-rotate) is as follows (according to equation 7 and having as coefficients the unknown variables X and Y instead of n and m):

$$X\theta' = X(\theta - \phi) = 2\pi - Y\phi$$

Since the parameter X is the only one to be multiplied by θ, the equivalence of the two equations (equation 5 and the above) dictates that $X = n$. From these two equations we get that $(X - Y)\phi = m\phi$, therefore $Y = n - m$.

Appendix 5

Part 1: Equations 36, 37 and 39 give us when co-rotating:

$$d = (nd + \phi)\theta/2\pi - (md + \theta)\phi/2\pi \Rightarrow n\theta - m\phi = 2\pi$$

which is equation 5, therefore equation 39 is correct. Similarly we get equation 40 from equations 37 and 38 when contra-rotating.

Part 2: Equation 37 gives us (branches co-rotating): $d = (I_\kappa 2\pi - \theta)/m$ and from equation 36 we get $I_\lambda = (n((I_\kappa 2\pi - \theta)/m) + \phi)/2\pi$. Therefore $I_\lambda = (n/m)I_\kappa - (n\theta - m\phi)/2m\pi$, which together with equation 5 gives us $nI_\kappa - mI_\lambda = 1$ (equation 41 with + sign at the right hand side). Similarly from equations 37, 38 and 7 we get equation 41 (contra-rotating). In the case of figure 5 (branches contra-rotate), the equations 37 and 38 give us the angle $d = {}^\wedge A_{0,3}SA_{5,0} = 137.5$degr by rotating clockwise.

If we rotate in the opposite direction we get $d' = {}^\wedge A_{0,3}SA_{5,0} = 360 - 137.5 = 222.5$degr, the equivalent 37 and 38 equations become: $2\pi I'_\kappa + \theta = md'$ and $2\pi I'_\lambda - \phi = nd'$. Since $I'_\kappa = m - I_\kappa$, $I'_\lambda = n - I_\lambda$, these give us $d' = I'_\lambda\theta + I'_\kappa\phi$ as equation 40 and $nI'_\kappa - mI'_\lambda = -1$ as equation 41 with the $-$ sign on the right hand side, related uniquely with the versions of equations 37 and 38 mentioned above.

In co-rotating branches we have equations 36 and 37 : $2\pi I'_\kappa + \theta = md'$ and $2\pi I'_\lambda + \phi = nd'$, so $nI'_\kappa - mI'_\lambda = -1$ (d' and d in opposite directions).

Part 3: From equation 42 we get: $r = (1/\lambda)^{I\lambda}\kappa^{I\kappa}$ which, when combined with equation 5 gives us:

$$r = \kappa^{-(m/n)I\lambda}\kappa^{I\kappa} = \kappa^{(nI\kappa - mI\lambda)/n}$$

and because of equations 41 and 5, this becomes $r = \kappa^{1/n} = \lambda^{1/m} < 1$.

Part 4: For spirals with co-rotating branches, if equations 6 and 8 are combined by equating their expressions, the following equation can be obtained:

$$\kappa^{(m+n)/n}\sin\theta - \kappa^{m/n}\sin(\theta + \phi) + \sin\phi = 0.$$

The above equation, (taking into account that $r = \kappa^{1/n}$), is equivalent to the equation 10, in Sushida, Hizume, Yamagishi [2, p.14].

Also, for spirals with contra-rotating branches, similarly to the above, we have:

$$\kappa^{m/n}\sin\varphi + \kappa\sin\theta - \sin(\theta+\varphi) = 0.$$

From equations 37 and 38, we get: $\sin(nd) = \sin\varphi$, $\sin(md) = -\sin\theta$, $\sin((m-n)d) = -\sin(\varphi+\theta)$, which together with $r = \kappa^{1/n}$, make the above equation equivalent to the equation 5 at [2, p.7].

References

1. H.S.M. Coxeter, *Introduction to Geometry*, John Wiley and Sons, Inc., 1989

2. T. Sushida, A. Hizume and Y. Yamagishi, *Triangular Spiral Tilings, journal of physics a: mathematical and theoretical*, 45, 2012 IOP Publishing

3. P. Przemyslaw Prusinkiewicz, A. Lindenmayer, *The Algorythmic Beauty of Plants*, Springer-Verlag, 1990

4. R. V. Jean, *Phyllotaxis, a Systemic Study in Plant Morphogenesis*, Cambridge University Press, 1994

5. Kevin Brown, Exponential Spiral Tilings,
   www.mathpages.com/home/kmath620/kmath620.htm

ALPHAMETICS

Charles Ashbacher

1. This problem is based on the "Star Trek: The Next Generation" episode number 35, "Measure of a Man." There is a trial where the purpose is to determine whether Data is a mere machine that can be deconstructed for study or is a sentient being with rights. Since the trial is a major one, we maximize the value of TRIAL.

```
   35
 WHAT
 DATA
   IS
_____

TRIAL
```

2. This problem is based on the first episode of "Star Trek: The Next Generation," "Encounter at Farpoint." The recurring character Q makes his first appearance and issues a challenge to Captain Picard and the crew of the Enterprise.

```
    1
    2
INTRO      Since the series was very popular, we maximize POINT
  TOQ
   AT
  FAR
_____

POINT
```

3. This problem is based on episode number 8 of "Star Trek: The Next Generation," "Justice." Wesley Crusher is sentenced to death for a minor crime on the planet Rubicon III.

```
    8
  WES
TODIE      Since the crime is so minor, minimize CRIME and
 FORA      solve in base 12
_____

CRIME
```

4. This problem is a salute to the actor that now plays Captain James T. Kirk in the Star Trek reboot.

```
TREK
PINE
  IS      we will of course maximize KIRK
 THE
 NEW
_____

KIRK
```

5. This problem is based on episode number 13 of "Star Trek: The Next Generation," "Datalore." Lore, the evil twin of the android Data, is introduced and we learn that they were both created by Dr. Soong.

```
  13
DATA
 AND    In this case, we will minimize LORE and solve in base 12
LORE
  BY
_____

SOONG
```

6. This problem is a douby-true alphametic in Western Aleut, a Native American language of the Aleutian Islands in Alaska and in Siberia. While it is not yet dead, there are now fewer than 500 native speakers.

```
ATAQAN        1
ATAQAN        1
ATAQAN        1
ATAQAN        1
_____     ____

SICHING       4
```

7. This problem is based on episode number 10 of "Star Trek: The Next Generation, Hide and Q." In that episode Riker is given the power of the Q and is suddenly omnipotent.

```
      10
   RIKER
   GIVEN   Of course we will maximize POWER and solve in base 12.
       Q
  _____

   POWER
```

BOOK REVIEWS

Edited by:Charles Ashbacher

Charles Ashbacher Technologies

5530 Kacena Ave

Marion, IA 52302

E-mail: cashbacher@yahoo.com

Viewpoints: Mathematical Perspective and Fractal Geometry in Art, by Marc Frantz and Annalisa Crannell, Princeton University Press, Princeton, New Jersey, 2011. 248 pp., $52.00 (hardbound). ISBN 9780691125923

This is a textbook/workbook that blends both art and mathematics while not skimping on either one. Tactics such as perspective and viewpoints are demonstated using both illustrations and the equations that describe them. It is designed for courses in mathematics for liberal arts, mathematics for artists and other interdisciplinary courses where art and mathematics are combined.

An artist's vignette follows each chapter and one of the common themes is that when the experienced artists are exposed to the mathematical explanations for the first time they recognize that the formulas represent what they have been doing all along. The mathematics is essentially applied geometry, there is a bit of algebra but it is all directly related to what appears in the drawings and images. This makes it much easier for the reader/student to understand the purpose and consequences of the equations.

The chapter titles are as follows:

*) Introduction to Perspective and Space Coordinates

*) Perspective by the Numbers

*) Vanishing Points and Viewpoints

*) Rectangles in One-Point Perspective

*) Two-Point Perspective

*) Three-Point Perspective and Beyond

*) Anamorphic Art

*) Introduction to Fractal Geometry

*) Fractal Dimension

Each chapter closes with a set of exercises and solutions to many are included in an appendix.

As can be seen from the content of this book, mathematics is the foundation of quality art, the discovery and application of perspective led to a dramatic change in the realistic nature of painting. This book is a textbook in the traditional format and is also a strong response to the question, "What is math used for anyway?"

Charles Ashbacher

Inlaid Magic Squares and Cubes, by John R. Hendricks, privately published, Victoria, B. C., Canada, 1999. 202 pp., ISBN 0968470017.

John R. Hendricks was one of the shining examples of a person dedicated to working in mathematics, constantly extending the field of magic squares. The magic square, an n-by-n grid of numbers such that all rows and columns add to the same value, has been a part of recreational mathematics for thousands of years.

Hendricks extends that out to magic squares within magic squares as well as magic cubes and diamond inlays. A magic cube is a three-dimensional structure of cells containing numbers where all the rows, columns and pillars have the same sum. Figure 1 is a deconstruction into layers of 3 x 3 x 3 magic cube.

Figure 1

1	17	24
15	19	8
26	6	10

Top

23	3	16
7	14	21
12	25	5

Middle

18	22	2
20	9	13
4	11	27

Bottom

With common sum 42.

Another form of magic square within a magic square is a diamond inlay. This is a rotated square within a magic square that is itself a magic square and Hendricks' first example of this appears in figure 2.

Figure 2

```
 9    7   14   20   15
16   21   24    3    1
 4    8   13   18   22
25   23    2    5   10
11    6   12   19   17
```

The diamond inlay is a 3 by 3 magic square with sum 39.

There are many very complex magic structures in this book and Hendricks is to be commended for a lot of truly creative work. He was a giant in the world of recreational mathematics and it is hoped that this review will help to better inform the mathematical world regarding his achievements.

Charles Ashbacher

Applications of Mathematics in Economics, edited by Warren Page, The Mathematical Association of America, Washington, D. C., 2013. 133 pp., $40.00 (paperback). Print ISBN 978-0-88385-192-0, Electronic ISBN 978-1-6144-317-9.

Simple and usable economic models can be developed using mathematics before calculus, they make excellent examples of how mathematics is used in the world. This book contains a set of nine fundamental economics principles that are introduced using the mathematics and sets of problems for the reader to consider and solve. Solutions to all of the problems are given at the ends of the chapters. The problems are not bunnies by any means, they are complex and descriptive of the material.

The chapter titles are a description of the coverage and they are:

*) Microeconomics

*) Scenarios involving marginal analysis

*) Intermediate macroeconomic theory

*) Closed linear systems

*) Mathematics in behavioral economics

*) Econometrics

*) The portfolio problem

*) Topics in modern finance

*) Maximizing profit with production constraints

As you can see from this list, there is significant breadth in the areas covered. Teachers of math courses from basic algebra to many math major courses will find problems that they can use in their classes.

Charles Ashbacher

Guns, Germs and Steel: The Fates of Human Societies, by Jared Diamond, W. W. Norton & Company, New York, New York, 2005. 528 pp., $29.95 (hardbound). ISBN 9780393061314.

This book is one of the most fascinating historical books that I have ever read, for it explains so much of what has happened when previously distinct societies have come into contact. In nearly all cases, one overwhelmed the other, often to the brink of extinction. Superior weapons and tactics are not the only explanation, it is a partial explanation but not enough to understand how a few hundred could militarily defeat tens of thousands of warriors in opposition. Naiveté on the part of the society about to be destroyed is also a partial explanation, but once again it is insufficient.

The primary reason is also the smallest of reasons, microscopic organisms that spread diseases that the conquerors were largely immune to while the conquered were not. There are recorded instances where the majority of the conquered society died within a short time of first exposure to the interlopers. This left them demoralized and their society on the verge of collapse, making them easy prey to the predatory humans.

While that is a fact that is generally known, Diamond traces the source back even further to ask and answer the most significant question, "Why was the transfer of deadly diseases so one-sided? Why were there not diseases in the Western Hemisphere or in Australia that were just as deadly to the Europeans?"

The answer is one that makes a great deal of logical sense, it is based on animal husbandry. Warm-blooded animals such as cattle, horses, sheep, goats and pigs were all domesticated in Eurasia and so there were many opportunities for microbes to make the leap from those animals to infect humans. This did not happen to the same extent in the societies that were destroyed. To this day the danger is well known, on occasion we hear the phrase "Bird Flu" and of course HIV is one virus that recently made the jump from other primates to humans.

Diamond also traces back the domestication of high yielding crops such as wheat in Mesopotamia and how it rapidly spread in a general east-west arc. This led to the organization of physically and politically stable societies with the organization and resources to build vast structures.

This is a fascinating book, one where the interpolation between the facts is clearly stated yet totally believable. Many societies have essentially been exterminated, more often by accident but often helped by malevolent human actions. This is also a look forward and a description of the dangers that still lurk from the actions of misguided microbes. It is in the best interests of microbes to kill humans very slowly or not at all so that they live long enough to transmit the microbes to other humans. Given the concentration of humans and the constant movement over short and vast distances, one microbe that mutates to be every infectious and lets humans live long enough to pass it along could easily kill millions in a very short time. The Spanish flu epidemic of 1918-1920 killed 3 to 5 percent of the world's population and was especially fatal for young, otherwise healthy people.

Charles Ashbacher

Quintillions, A Set of Solid Pentominoes created by Kadon Enterprises
http://www.gamepuzzles.com/

This is a set of solid pentominoes where the squares are approximately ¾ of an inch thick. The craftmanship is extraordinary, all of the edges and faces are as smooth as possible and the grain of the wood is very attractive.

The set comes with a folding grid for the pentominoes that is 9 by 12, giving the "player" great flexibility in how they put the pieces together. The package contains a 63 page booklet of structures that can be made from the set. As is normally the case, there is a wide difference in the difficulty of the problems, some are three-dimensional structures.

There are hours of challenging fun inherent in this collection, the pentominoes are the ideal set of polyominoes, there are enough of them to make it interesting but not so many as to be impossible. While few toys can double as a work of art, this is definitely one of them.

Charles Ashbacher

L. A. Math: Romance, Crime and Mathematics in the City of Angels, by James D. Stein, Princeton University Press, Princeton, New Jersey, 2016. 256 pp., $24.95 (hardbound). ISBN 9780691168289.

This is a book that teachers of mathematics from middle school through college will find useful. Stein uses the always interesting private investigator context to present a series of 14 problems in basic mathematics. Each is wrapped within a reasonable and believable scenario that all people will be able to understand.

As is usually the case with investigators there is a main character with a sidekick, Freddy Carmichael is an investigator and Pete Lennox is the sports addicted sidekick that knows a great deal of mathematics. In general Freddy takes the case, hears the situation and then Pete asks one or two critical questions and provides the solution with mathematical justification.

Topics of the investigations include what is known as the Monty Hall problem, why going up 20% and then down 20% does not get you back to start, why going 40 mph one way and 60 mph the other way is not an average of 50 mph overall, the consequences of compound interest, basic game theory and fundamental counting principles. A small, separate appendix containing additional information about the underlying mathematics is included for each of the 14 problems.

With the math within reach and the presentation wrapped in a bit of entertainment, these problems are both a way to make math learning fun and also a partial answer to the common question, "What is math used for?"

Charles Ashbacher

How Euler Did It, by C. Edward Sandifer, The Mathematical Association of America, Washington, D. C., 2007. 304 pp., $51.95 (Hardbound). ISBN 978-0-88385-563-8.

Leonhard Euler was arguably the most prolific mathematician of all time, the breadth of his coverage is the most impressive aspect of his work. He literally created several new areas of mathematics and mathematicians continue to expand and refine his work.

Yet, there is still the question of how he actually proved the results that so many mathematicians have committed to memory. This book is a collection of annotated reproductions of some of the most memorable of Euler's results. The 40 items all appeared in the column "How Euler Did It" that was published in the MAAOnline column between November, 2003 and February, 2007.

The columns are organized into four groups: geometry, number theory, combinatorics and analysis. Even if you were already astounded at Euler's accomplishments, that emotional state will be expanded even further. For in reading these columns, you are often faced with the question, "How did Euler ever think to do that?"

When you are faced with the question in the title, the simplistic response is "Euler was a genius in math." True and obvious, but not descriptive. In this book you are privileged to see that genius in action, where you see some of the thought processes that led to the "Voila" moment in the reader when the proof is complete.

Charles Ashbacher

Creative Mathematics, by Alan F. Beardon, Cambridge University Press, New York, New York, 2009. 122 pp., $34.99 (paperback). ISBN 9780521130592.

This book contains a series of problems created and refined by the author in a course on problem solving given at the African Institute for Mathematical Sciences in South Africa (AIMS). The institute accepts graduate students from all over Africa.

There are 11 problems, the section containing a listing of all the problems comes first, followed by the complete set of solutions. The coverage is rather broad, some can be considered to be puzzles rather than problems. For example, problem D is about tetrahedral dice, problem F involves triominoes and problem G involves a set of weights $\{1, 2, 3, \ldots, n\}$ and the question is, "How many weights can be discarded while retaining enough to weigh all values from 1 through n?"

The problems are not that complicated in the sense that they require an in-depth knowledge of mathematics in order to derive the solution. For some of them the author openly suggests the reader write a simple computer program in an attempt to solve it.

The real point of the problems is to see the solutions and experience the insight needed to arrive at the solution. For example, the solution to problem G is immediately obvious if you think in binary and use the weights that are the powers of two less than or equal to n.

As a math book, this one is ordinary regarding the mathematical content. The real value is in the insights that are invoked in creating the solution.

Charles Ashbacher

Enjoy Puzzling With biLLies, by Paulus Gerdes, Lulu Publishing, Maputo, Mozambique, 2009. 252 pp. $23.08 (paperback), $5.95 (downloadable PDF). ISBN 978-0557166961.

The set of polyominoes called biLLies are formed by the attachment of two L-triominoes together. There are 14 biLLies, so the combinatorics favor the creation of a large number of construction problems.

There are 24 chapters, each of which opens with a set of figures designed to be constructed using a subset of the biLLies. Solutions to all then follow in the same chapter. In all cases the author is very explicit in stating that the solutions are only a sample of the possibilities rather than a complete set.

If you are interested in polyomino constructions, this is a book that can keep you occupied for quite some time.

Charles Ashbacher

Expeditions in Mathematics, edited by Tatiana Shubin, David F. Hayes and Gerald L. Alexanderson, The Mathematical Association of America, Washington, D. C., 2011. 400 pp., (hardbound). ISBN 9780883855713.

Alternating on the campuses of San Jose State University and Santa Clara University in the center of the area of California known as Silicon Valley, there was a series of bi-monthly mathematical lectures called the Bay Area Mathematical Adventures (BAMA). The lectures are designed for the general audience, although there is no sparing of the mathematics. People at all levels of mathematical knowledge attend and are engaged rather than passive. The lecturers are generally nationally and internationally known mathematicians, describing their work. This book is a collection of their presentations.

 The breadth of the coverage is best described by citing the categories that appear in the table of contents. They are:

*) General, covering basic mathematical paradoxes, Sudoku, cardinal numbers and tricks with cards.

*) Number theory

*) Geometry & topology

*) Combinatorics & topology

*) Applied mathematics of observing the sun and moon from different locations on the Earth, zero knowledge proofs and the value of combination drug therapies.

 The papers are all fascinating to read, even those not within your general areas of mathematical interest. My favorite single point was an example of Simpson's Paradox where an improperly interpreted data set can indicate gender bias when no such bias exists. This is popular mathematics at the absolute best, understandable with no lack of mathematical rigor.

Charles Ashbacher

Single Digits: In Praise of Small Numbers, by Marc Chamberland, Princeton University Press, Princeton, New Jersey, 2015. 240 pp., $26.95 (hardbound). ISBN 9781400865697.

 To be more precise, the numbers examined in this book are the counting numbers 1 through 9. Chamberland takes the reader through many different aspects of the use of these numbers, with more pages devoted to the lower numbers. For example, there are 23 pages covering "1," 45 for "2," 14 for "8" and 11 for "9."

 There is great breadth in the topics covered, number theory, tilings, packing, dissections, chaos, coding, graph theory and fractals are some of the topics. All are presented in a manner typical of popular mathematics, it takes some mathematical background to understand them, but nothing the follower of popular mathematics will find overwhelming.

There is also no continuity of the story line, one can go to the first page of any of the chapters and begin reading with no lack of context. The chapter title tells you all you need to know about the subject, various ways that that specific number appears in operations.

This is a sound book of popular mathematics, Chamberland covers a lot, in enough background to generate your interest but not so much to overwhelm with detail. He also never shirks from using a formula where one is needed. In a world where some authors lack the courage to use complex formulas, Chamberland stands taller than many of his contemporaries.

Charles Ashbacher

Problems and Conjectures

Edited by: Lamarr Widmer

Readers are invited to send new problems, solutions and comments to me at *Messiah College, Suite 3041, One College Avenue, Mechanicsburg, PA 17055* or email to widmer@messiah.edu . Put each problem or solution, with your full name and postal address, on a separate sheet. Selection of solutions for publication will take place at least three months after problems appear in print.

1. A Topical Conversation by Andy Pepperdine, Bath, UK

A short time ago, Jim was talking to his neighbor, Zoe on a typical British street with even numbered houses in order on one side and odd numbers on the other.
"Do you realize that if I multiply my age to the nearest year by my house number, then I get this year's number, in the Gregorian calendar of course?"
"Really? Well, in that case, at this time next year, multiplying your age by my house number will get the number of the year", replied Zoe.
In what year did this conversation take place, how old was Jim and at which numbers did he and his neighbor live?
When was the previous occasion this conversation could have taken place? When will the next such time be?

2. Squaring the circle by Hubert Hagadorn, Menlo Park, CA

A circle is cut into six pieces that are fitted into a unit square with no overlapping. What is the area of the largest circle for which this can be accomplished and how should it be cut?

3. Presence of All Ten Digits by Hubert Hagadorn, Menlo Park, CA

a. What is the minimum number of digits that a number must have so that there is at least a 50% probability that all digits 0 through 9 are present?
b. What is the precise probability when the number of digits is as in part a?

4. Another Triangular Inequality by Henry Ibstedt, Issy les Moulineaux, France

Prove that the following inequality holds in any triangle ABC , with sides a, b and c, inscribed radius r, and radii r_a, r_b and r_c of the circles tangent to one side and the extensions of the other two.

$$\frac{abc}{r} \geq \frac{a^3}{r_a} + \frac{b^3}{r_b} + \frac{c^3}{r_c}$$

5. An Attractive Triangle Relation by Henry Ibstedt, Issy les Moulineaux, France

Prove that the following relation holds in any triangle ABC, with sides a, b and c and area T.

$$4T = \frac{a^2 + b^2 + c^2}{\cot A + \cot B + \cot C}$$

SOLUTIONS TO ALPHAMETICS

1.

```
     35
   9401     with 9 and 8 interchangeable
   8010
     56
  _____

  17502
```

2.

```
      1
      2
  79640
    603
     26
    524
  _____

  80796
```

3.

```
               8
        10   4  11     with S & A and W & D interchangeable
   1   6  8   0   4
       7  6   3   5
  _____

   2   3  0   9   4
```

4.

```
   3469
   5716
     78
    326
    160
  _____

   9749
```

5.

```
            1   3
   8  11  10  11
      11   9   8
   2   0   4   3
              7   5
   ─────────────────
1   0   0   9   6
```

6.

```
393632
393632
393632
393632
──────
1574528
```

7.

```
                10
   6   4   1  10   6
   5   4   0  10   9        with 9 & 3 and 1 and 0 interchangable
                  3
   ──────────────────
  11   8   2  10   6
```

Row by row—from Latin Squares to Tic-Tac-Toe

by Kate Jones

The theme of lining things up and sorting them by type is very ancient.

An early riddle challenged puzzlers to arrange figures representing military officers in a 6 x 6 square grid so that every space held one piece and every row, vertically and horizontally, contained 6 different levels of officer and one each from 6 different regiments. It is possible to achieve such a formation, but only orthogonally, not diagonally. The 6 x 6 is stubborn that way. You'll have much better luck with a 4 x 4 or 5 x 5 square, where the long diagonals can also be all different.

Such arrangements are known as "Latin Squares", reportedly from versions solved with different letters of the Latin alphabet as explored by Leonhard Euler in the 18th century. Wikipedia has a math-heavy report and analysis of Latin Squares for those who are looking for an in-depth study of them here: https://en.wikipedia.org/wiki/Latin_square.

The idea of forming straight lines with various design conditions has led to historical games like Tic-Tac-Toe, Nine Men's Morris (also made by Kadon in a nice handcrafted wood set), Bingo, and the latter-day phenomenon, Sudoku. A special relative of this genre is the "Magic Square" where all-different numbers form rows with all equal sums.

Let's look at a few samples from this rich repertoire of game and puzzle ideas that play with pieces lined up by certain rules. Kadon as of 2015 has 54 original recreational math sets that incorporate challenges of organizational distribution. They are all shown here: http://www.gamepuzzles.com/tictacto.htm

Let's look at a few samples from this rich repertoire of game and puzzle ideas that play with pieces lined up by certain rules. Kadon as of 2015 has 54 original recreational math sets that incorporate challenges of organizational distribution. They are all shown here: http://www.gamepuzzles.com/tictacto.htm

Rows of 3

An amazingly complex meta-game, **Proteus**, has changing rules for moving, trading tiles, and winning. One of the goals

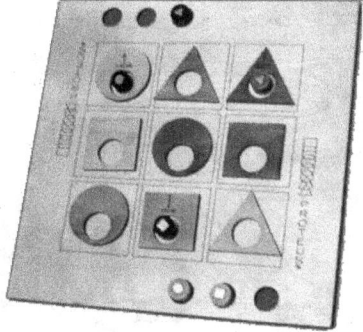

is to get three men in a straight line. Tiles on the board can have 9! (9-factorial) placements. The arrangement at left has 3 different shapes and 3 different colors in every row and column. Invented by Michael Waitsman.

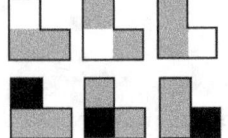 In *Pocket Vees*, two players compete to form a straight line with their 3 squares (black or white) while the Vee tiles interlink and block. Invented by Kate Jones.

With *Six Disks,* two players stack their 3 disks three layers high in Vertical TicTacToe. Win by getting your disks in one stack or across the top row. By Kate Jones.

 In *RunnuRound*, up to four players race to find a sequence of three numbers in a scrambled panel of four differently colored rows, each with 10 digits. Sequences can "wrap around" at row ends and go forward and back. Invented by Joe Marasco.

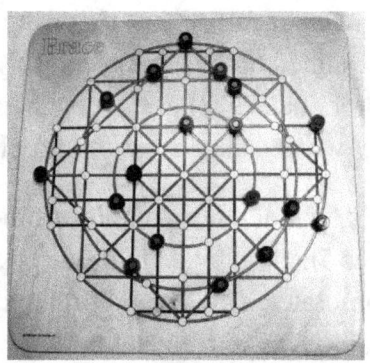

Brace, by Alan Kross-Vinson, is a strategy game for two where pieces move on color-coded paths to surround one player's piece by two of the other: an embrace (not a capture) along a straight or curved line that scores points. A lovely idea, and a challenge to form multiple rows of 3 in one move.

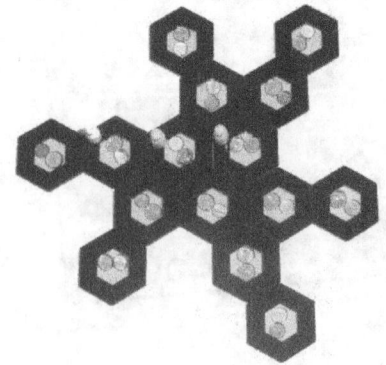 In *Gemstones*, up to five players compete to collect different colors of gems from three "mines" in a straight line. Colors have varying point values to add up to the winning score. Watch out

for the guardians of the mines—refill what you plundered. Invented by P. R. Chase.

Rows of 4

Our purest 4x4 Latin Square is **Bear Hugs Jr.**, a set of 16 jolly little teddy bears with 4 different colors, 4 different arm positions and 4 different leg positions in every row and column. This solution in addition displays the Sudoku feature of a maximum of 7 embedded and overlapping 2x2 subregions that also have 4 different colors/arms/legs. Because 6 pairs of mirror-image "twins" can be switched, the bears can fit their cut-out spaces in 64 different ways, one of which has solid-color columns.

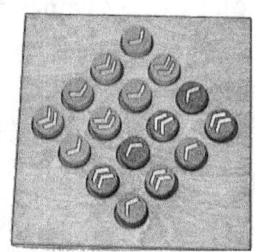

A 4x4 gameboard, **Void**, invented by Michael Waitsman, includes a "Row of Four" strategy game where players win if their pawns, moving like chess kings, form a line of four of the same color or four of the same symbol (single or double arrows). A more complex goal is to create an empty row—a row of four "voids".

A+d+d+d is a clever strategy game by Devin Stewart where two players roll 5 dice and use one or all of the numbers to add up to a number on the board on which to place their pawns. Get 3 or 4 or even 5 pawns in a row to build your score.

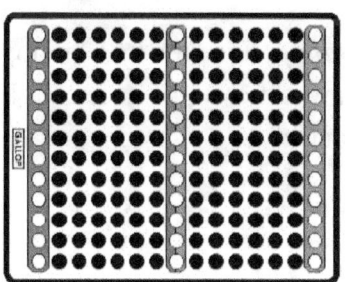

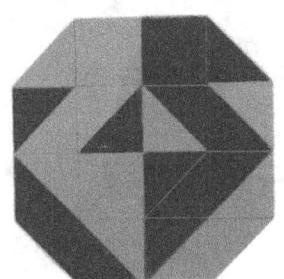

Triangoes Jr. has 15 tiles--triangles, squares, parallelograms left and right—in every mix of two colors. It makes endless beautiful patterns (left) and plays a form-a-row strategy game

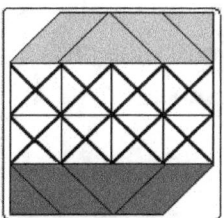

121

(right) where tiles move as a transforming connected group within a 4x4 grid to the opposite side, a race to the finish line.

Instant Insanity is the famous old puzzle with 4 multi-color cubes that are to fit into a row of 4 with all four sides displaying four different colors. This one is inventor Frank Armbruster's new, trickier revision.

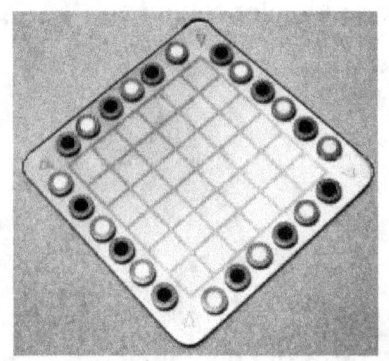

Hmm... is one of an anthology of 32 games played on the ***Six by Six*** gameboard. Two players enter pieces rook-wise from the edge along rows to get four in a straight line horizontally, vertically, or diagonally. Once inside, pieces move only one rook step at a time. Invented by Christopher Clark. This board also serves for Magic Square sizes 3x3, 4x4, 5x5 and 6x6, arranged with disks numbered 1 to 36. A further compendium of games and puzzles invented by Stephen Sniderman uses a 5x5 section of the board. On the full board also play Numerical Criss-Cross, invented by Kate Jones, where rows of numbers form equations.

Gallop, invented by Christoper Clark, also includes, among its five games, a strategy goal where pawns move like queens to line up 4 in a row anywhere on the 12x13 field. Developed by Kate Jones

Rows of 5

Among the games for the complex ***End Point*** gameboard is ***End Jam***, where players seek to fill up the other player's five end row points while pieces move along arcs and loops. Invented by Arthur Blumberg.

One of the games in the **Octiles** set, for up to four players, is **Team Up**, where a player's 5 pawns are to connect up serially by moving on changing paths. A challenging solitaire asks that all the tiles form the longest single uninterrupted line, using even any of the exterior loops on the board. Invented by Dale Walton.

Colormaze, on its 8x8 board, includes a whole series of Latin Squares formed with colors, especially the 5x5 with wrap-around features. Another solitaire theme is forming rows of all different colors on the four borders, unraveling them from a maze of colors packed at the center, advancing from 2 to 6 colors. Developed by Kate Jones.

Rows of 6 or more

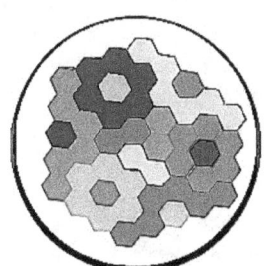

Hexdominoes with 21 domino-like tiles in every combination of 6 colors offers the widest assortment of shapes to solve with no repeating color in any row, provided no row is longer than 6 cells. A hexagon grid has rows in all 3 directions, and all its rows count. Or arrange to have 12 rows of 4 matched colors (right).

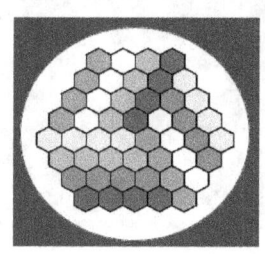

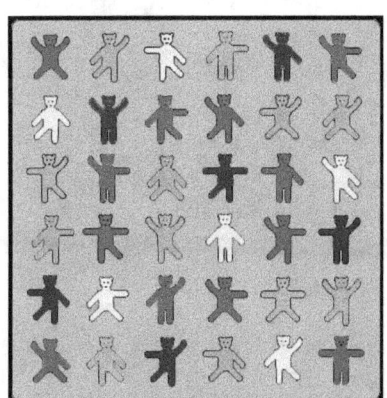

The 6x6 **Bear Hugs** has 36 all-different poses, with 6 colors and 6 arm positions in each horizontal and vertical row, like the

123

classic regiment/officer Latin Square. In addition, this solution contains 18 rectangles, 2x3, with 6 different colors, à la Sudoku. Further, the diagonal from upper left to lower right has the six symmetrical bears. All others constitute 15 pairs of mirror-image twins that can be swapped, mixing colors in 2^15 ways.

 **Roundominoes** is not just a puzzle but also several games. In one called **Ringo-Bingo**, the players seek to fill a row of 7 spaces, horizontally, vertically or diagonallly. The convoluted pieces can surprise by how they fit.

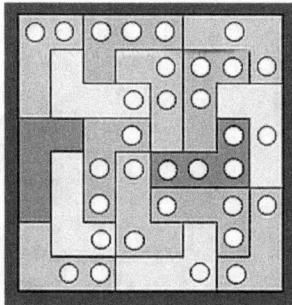 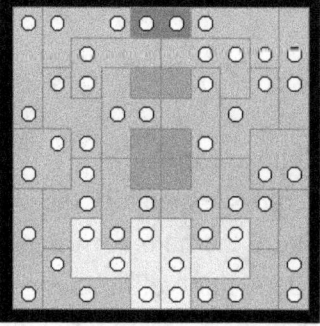

The 8x8 **L-Sixteen** Nevzat Moraç and Kate Jones) and 10x10 Kate Jones put holes in respectively, in all (proposed by developed by **Fill-Agree** by rows, 4 and 5 three directions—

horizontal, vertical, diagonal. Many other tasks are included for lining them up.

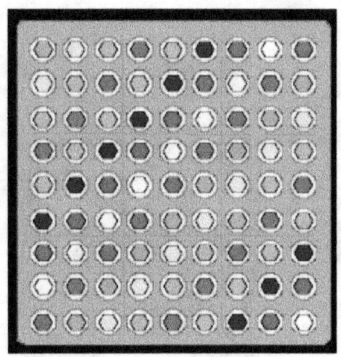 **Pseudo-Coup** is the classic 9x9 version of Sudoku with a twist of colors instead of numbers. Playable also as several non-match strategy games, including three by Henry Kwok.

Finally, **11 Magic Cubes**—a Magic theme with numbers 1 through 6 on the 11 hexominoes that can fold into a cube, designed by Kate Jones. Place numbers in cells so all rows in a figure have the same sum:

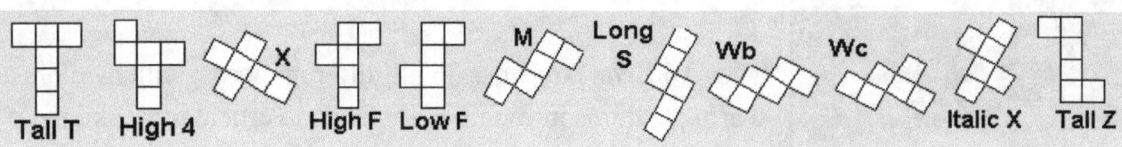

Tall T   High 4   High F Low F   M   Long S   Wb   Wc   X   Italic X   Tall Z

After filling in your solutions on a large copy of these shapes, cut out and fold them into cubes and tape their edges closed. Arrange the 11 cubes into the 7 figures below so that all rows—

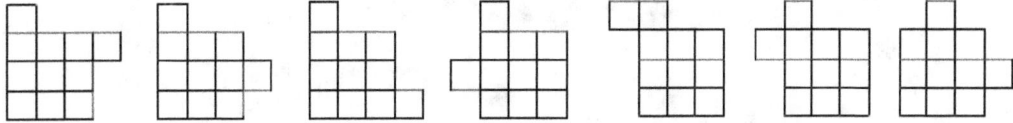

north-south and east-west and the vertical walls on all four sides—give the same sum.

Next issue: "Making Connections" — the antidote to chaos.

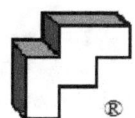

Mathematical Spectrum

*A magazine for students
and teachers of mathematics
in schools, colleges and universities*

Editor: D. W. Sharpe, *University of Sheffield*

For over three decades, *Mathematical Spectrum* has been a popular source of stimulating ideas for teachers, students and mathematical enthusiasts alike. Articles cover a wide range of topics in mathematics and the related sciences as well as the history of mathematics, with regular education and computer columns, a letters page, problems and solutions, and reviews of books and software.

Contributors from all over the world include established mathematicians as well as students — we welcome original student contributions and award annual prizes for the best ones published.

Subscription information: Vol. 47 (Sept 14–Aug 15) $24.50; Vols 47 and 48 (Sept 14–Aug 16) $47.00; Vols 47, 48 and 49 (Sept 14–Aug 17) $67.50. Three issues per volume in September, January and May; postage and handling included. To subscribe, contact:

Mathematical Spectrum   Tel: +44 114 222 3922
Applied Probability Trust   Fax: +44 114 222 3926
School of Mathematics and Statistics
The University of Sheffield   Email: s.c.boyles@shef.ac.uk
Sheffield S3 7RH, UK   Web: www.appliedprobability.org

Published by the **Applied Probability Trust**, a non-profit-making organisation based in the University of Sheffield

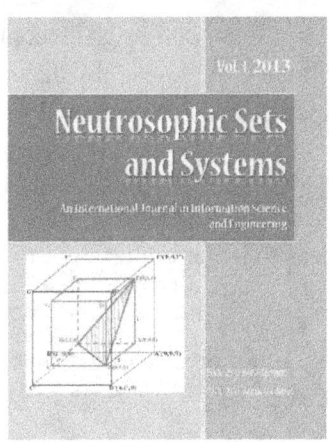

Editor-in-Chief:

Prof. Florentin Smarandache

Department of Mathematics and
Science

University of New Mexico

705 Gurley Avenue

Gallup, NM 87301, USA

E-mail: smarand@unm.edu

Home page:
http://fs.gallup.unm.edu/NSS

Associate Editors:

Dmitri Rabounski and Larissa Borissova, independent
researchers.

Said Broumi, Univ. of Hassan II Mohammedia, Casablanca,
Morocco.

A. A. Salama, Faculty of Science, Port Said University, Egypt.

Yanhui Guo, School of Science, St. Thomas University,
Miami, USA.

Francisco Gallego Lupiañez, Universidad Complutense,
Madrid, Spain.

Peide Liu, Shandong Universituy of Finance and Economics,
China.

Pabitra Kumar Maji, Math Department, K. N. University, WB,
India.

S. A. Albolwi, King Abdulaziz Univ., Jeddah, Saudi Arabia.

Mohamed Eisa, Dept. of Computer Science, Port Said Univ.,
Egypt.

Neutrosophic Sets and Systems has been created for publications on advanced studies in neutrosophy, neutrosophic set, neutrosophic logic, neutrosophic probability, neutrosophic statistics that started in 1995 and their applications in any field, such as the neutrosophic structures developed in algebra, geometry, topology, etc.

The submitted papers should be professional, in good English, containing a brief review of a problem and obtained results. Neutrosophy is a new branch of philosophy that studies the origin, nature, and scope of neutralities, as well as their interactions with different ideational spectra.

This theory considers every notion or idea <A> together with its opposite or negation <antiA> and with their spectrum of neutralities <neutA> in between them (i.e. notions or ideas supporting neither <A> nor <antiA>). The <neutA> and <antiA> ideas together are referred to as <nonA>.

Neutrosophic Set and Logic are generalizations of the fuzzy set and respectively fuzzy logic (especially of intuitionistic fuzzy set and respectively intuitionistic fuzzy logic). In neutrosophic logic a proposition has a degree of truth (T), a degree of indeter

minacy (I), and a degree of falsity (F), where T, I, F are standard or non-standard subsets of $]^{-}0, 1^{+}[$.

Neutrosophic Probability is a generalization of the classical probability and imprecise probability.

Neutrosophic Statistics is a generalization of the classical statistics.

What distinguishes the neutrosophics from other fields is the <neutA>, which means neither <A> nor <antiA>.

<neutA>, which of course depends on <A>, can be indeterminacy, neutrality, tie game, unknown, contradiction, ignorance, imprecision, etc.

All submissions should be designed in MS Word format using our template file:

http://fs.gallup.unm.edu/NSS/NSS-paper-template.doc

A variety of scientific books in many languages can be downloaded freely from the Digital Library of Science:

http://fs.gallup.unm.edu/eBooks-otherformats.htm

To submit a paper, mail the file to the Editor-in-Chief. To order printed issues, contact the Editor-in-Chief. This journal is non-commercial, academic edition. It is printed from private donations.

Information about the neutrosophics you get from the UNM website:

http://fs.gallup.unm.edu/neutrosophy.htm

The home page of the journal is accessed on

http://fs.gallup.unm.edu/NSS

BOOKS IN RECREATIONAL MATHEMATICS BY CHARLES ASHBACHER AND ASSOCIATES

Topics in Recreational Mathematics 1/2015 ISBN 978-1507603215

Topics in Recreational Mathematics 2/2015 ISBN 978-1508617099

Topics in Recreational Mathematics 3/2015 ISBN 978-1511641005

Topics in Recreational Mathematics 4/2015 ISBN 978-1514317518

Topics in Recreational Mathemastics 5/2015 ISBN 978-1519115676

Alphametics as Expressed in Recreational Mathematics Magazine ISBN 978-1508538134

Ten Year Cumulative Index to the Journal of Recreational Mathematics, edited by Joseph S. Madachy and Charles Ashbacher ISBN 978-1508936800

Alphametics Expressing Thoughts From the Star Trek Original Series ISBN 978-1512152784

Mathematical Cartoons ISBN 978-1514207130

Solved Problems in Statistical Inference ISBN 978-1515215622

Associates

Artist Catie Ribble

Editor Rachel Pollari

Editor Jennifer Corrigan

Artist Jenna Richardson

www.ingramcontent.com/pod-product-compliance
Lightning Source LLC
Chambersburg PA
CBHW081607200526
45169CB00021B/2210